The
People's
Power

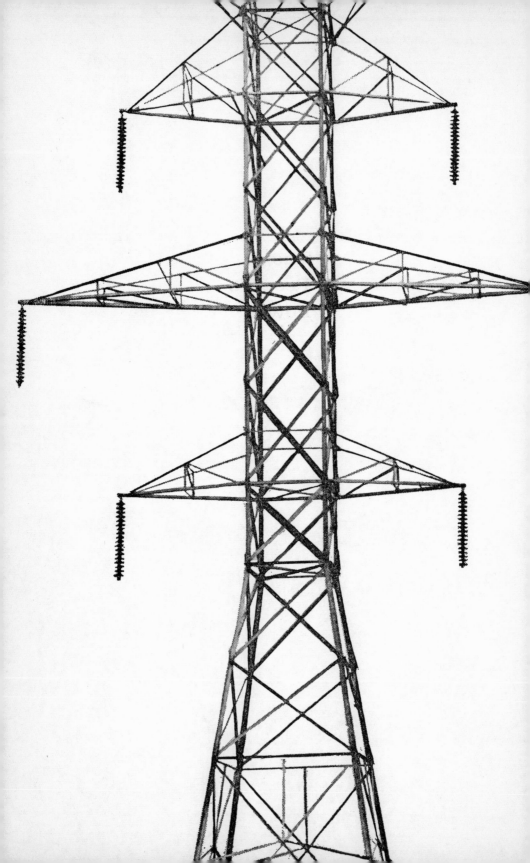

MERRILL DENISON # The People's Power

THE HISTORY OF ONTARIO HYDRO

MCCLELLAND & STEWART LIMITED

DESIGN

Jacket Hans Kleefeld MTDC
Portfolio Oscar Ross MTDC

PRINTED AND BOUND IN CANADA
T. H. Best Printing Company, Ltd.

CONTENTS

v

ILLUSTRATIONS

FOREWORD

As a boy I grew up in an Ontario family in which the concept
of public ownership was supported with passionate concern,
and as a student at the University of Toronto I followed with
intense interest Ontario Hydro's early struggles to become securely
established and vindicate the principles that brought it into being. At
a later date, as secretary of the American Scenic and Historic Preservation Society, whose founders initiated the movement to preserve the
beauty of Niagara Falls, I became familiar with the story of the great
river, and as an industrial historian, at a still later date, with the history
of electrical technology and development.

Notwithstanding such preparations, to recapture and interpret the
history of Ontario Hydro has proved more difficult than could have
been anticipated. Political climates become more temperate in retrospect; the meanings of words have changed; the memories of aging
stalwarts have become faulty and their recollection of bygone events
conflicting. For these reasons, the work has been based entirely on
documentary material, principally the Annual Reports of the Commission to the Legislature of Ontario, contemporary newspaper files,
and other references in the voluminous archives of the Commission.
This material is so readily available and so extensive and diverse in
character as to make formal annotation redundant.

I have also received great help from W. R. Plewman's *Adam Beck
and the Ontario Hydro*, published by the Ryerson Press, Toronto,
1947. Without this biography of the first Chairman of the Commission, based on Mr. Plewman's newspaper experience and long journal-
istic association with Sir Adam, some of the early events in the Hydro
story could not have been traced.

Another author to whom I owe much is Ernest Gruening, now
United States Senator for Alaska, whose *The Public Pays*, published
by the Vanguard Press, New York, 1933, presents an analysis of the

vii

massive findings of the United States Federal Trade Commission's investigation into public electrical utilities, begun in 1929, and has saved me untold labor. I am also indebted to Professor John H. Dales, of the Department of Economics of the University of Toronto, for personal advice and excellent work on electrical development in Quebec, *Hydroelectricity and Industrial Development, Quebec, 1898-1940*. I have also found helpful *Niagara in Politics* by Professor James Mavor; *Hydro Electric Development in Ontario* by E. B. Biggar; *Toronto Hydro Recollections* by E. M. Ashworth; and *The St. Lawrence Seaway* by Lionel Chevrier, to whom acknowledgments are extended.

On a more personal plane I wish to extend my thanks to Dr. Richard L. Hearn and to Mr. James S. Duncan, the Chairmen of the Commission under whom *The People's Power* was begun and completed, for their unfailing coöperation, and to all those others of the Hydro organization, at headquarters and in the field, who responded so helpfully to demands made upon their time.

Several persons have been associated with the work in its various stages of development, among whom I wish particularly to thank John Coulter, James Rorty, and P. F. O'Dwyer. Thanks are also extended to Paul Clark for contributing original background material on early electrical developments in Canada, and to E. C. Ertl, Mrs. Graham Spry, and John Martin for their constructive criticisms of the manuscript.

Photographs reproduced are by Ontario Hydro, except where otherwise noted.

Merrill Denison
Bon Echo
October 1, 1960

THE SEED BED

1

The Hydro-Electric Power Commission of Ontario has long been acclaimed by its friends and enemies alike as one of the most successful enterprises of its kind to be found anywhere in the world: by its friends, for more than half a century, through their unwavering support of the pioneer Canadian publicly owned electrical utility; by its enemies through their unremitting attacks on its philosophy of ownership and its management policies over the same period of time. Its methodology and accomplishments arouse little controversy now in Canada but are still attacked in the United States.

Hydro—as the Commission and all its works are familiarly called —is no longer a unique institution. Several other Canadian provinces now have similar hydro-electric power commissions, and its statutory structure and management techniques have been copied in the United States and other countries. Nevertheless, Hydro was unique at its inception. Having its beginning when the technologies of electrical production and transmission were in their infancy, the Commission was one of the world pioneers in large-scale hydro-electric development, standardization of equipment and appliances, utility management and rural electrification.

Established in a climate of evangelical fervor early in the present century, the enterprise was brought into being through a series of legislative enactments which enabled a small number of municipalities in the western part of the province to coöperate in financing the purchase, transmission, distribution, and sale at cost of electrical energy generated at Niagara Falls, approximately a hundred miles away. With an initial capital of $2,500,000 advanced by the Provincial Government, the Commission by 1910 had built its first transmission line and begun to deliver power to eight municipalities which had by that date voted to join the public-ownership system.

From this small beginning Hydro has grown in just over fifty years

1

to become one of the largest single producers and distributors of electricity in the world. Its investments in plant, structure and other facilities now total more than $2,500,000,000 and its dependable capacity has grown to more than 6,000,000-kw. With the exception of 2,000,000-kw in Northern Ontario, every hydraulic potential in Ontario has been developed, including those of the Niagara and St. Lawrence Rivers. Hydro's transmission lines now form a series of networks carrying energy to every settled part of a province stretching approximately one thousand miles from south to north and from east to west. The system had grown by the end of 1959 to include 354 municipal utilities, more than 47,000 miles of rural lines serving 94 per cent of Ontario's farms and, in addition, direct sales to large consumers of electricity in the mining and pulp-and-paper industries of Northern Ontario. Apart from provincial grants to help meet the capital costs of rural electrification, Hydro has paid its way from the beginning at low cost to the consumer and at no cost to its original backers: the ratepayers of the coöperating municipalities.

Statistics can bring into high relief the physical growth of Hydro during its relatively brief life span but they reveal little or nothing of its organic qualities or the economic and social changes it has effected. The economic impact has been no less than stupendous. In a very real sense, Hydro has made modern industrial Ontario, of all ten Canadian provinces, the most populous and the richest in terms of gross production. In conjunction with the internal combustion engine, Hydro has wrought an equally beneficent cultural revolution, particularly in the so-called "backwoods" areas where farmlands give way to a rocky tree-covered wilderness with streams, lakes and glacial overburden, unsuited even for subsistence farming. There, even after the end of World War I, people were living much as they had lived in the eighteenth century, with atrocious roads and few, if any, of the amenities of modern civilization. In this rocky, lake-spattered area, abandoned a generation earlier by the lumbering industry, Hydro, in conjunction with its coeval technological twin, the self-propelled vehicle, has wrought a modern miracle of social regeneration.

The introduction of electricity by private enterprise has, of course, brought similar economic and social benefits elsewhere, though usually at a greater cost to the consumer than in Ontario under public ownership. The question arises, therefore: what impelled one Canadian province among all the states of North America—or elsewhere for that matter—to embark on such a novel experiment?

The answer is to be found in the varied influences—geological, geographic, demographic, economic, social, cultural, political and historical—which combined to mold the Ontario of the 1890's and early 1900's—in short, in the ecological environment. Although more exposed to American influences than any other part of Canada, the province was not wholly Americanized: though the most richly endowed and most centrally located of the Canadian provinces, neither was it typically Canadian. These paradoxes combined to produce the soil in which the boldly planted seeds of social experiment and change could germinate and grow to bear abundant fruit.

To understand the peculiar nature of that soil, however, it is necessary first to understand the influences that produced it. The most important of these were geographical and historical, for these together produced the psychological climate that led to the Hydro experiment in public ownership. As always, the basic factor was the land, for it is the land that provides the people with their economic sustenance, molds their character, determines their social and political outlook, and largely determines the course their history shall pursue.

One outstanding fact about the Canadian land is that it forced Canada to remain one of the "have-not" countries until the beginning of the present century. Now one of the richest on the globe, with still unrealized potentials, it did not become so until science and technology were able to provide the tools to develop its strikingly extensive natural resources.

The explanation is to be found in the Pre-Cambrian Shield, a vast wedge-shaped massif of ancient crystalline rocks comprising nearly half of Canada's land surface: no less than 1,850,000 square miles. Formed during the convulsive eons of the cooling of the planet's crust, this warped and eroded hinterland extends from the Canadian-U.S. boundary on the St. Lawrence River northward to beyond the Arctic Circle, from the Canadian prairies eastward to the Atlantic Ocean, thus forming a 1,200 mile barrier between Central and Western Canada. The rocky, scarred plateau—known to geologists as a peneplain— bearing still the marks of its molten origin, denied the country a "middle west" comparable to that in the United States and retarded its population growth for decades. Nowhere in the world, except possibly in the great deserts, has man encountered a more inhospitable region for settlement or a more difficult area for economic development. Scraped to the bone by glacial action in recent geological time, the Shield provided no agricultural land except where lean podsols lie trapped between the folded hills and no easily exploitable resources except timber. While the region supported deep successive bands of forest growth, these became progressively less valuable to the north, to vanish finally where the ground is permanently frozen to a depth of hundreds of feet. Locked in the ground were immense stores of mineral wealth, but these had to await both discovery and the ability to extract them.

In one respect only had the Shield seemed kind to Canada. In their advance, successive ice masses had bulldozed its surface, altering the course of waterways and obliterating the beds of ancient rivers; in melting and decay, the same glaciers had left exposed a vast new reservoir system of rivers, lakes and catch basins which, together, give Canada 48 per cent of the world's supply of fresh water and more fresh-water lakes than all the rest of the countries of the world combined. The same ponded waters seeking outlets to the sea by innumerable water courses, provided Ontario and Quebec with hydraulic resources of 28,000,000-hp. It was this great power potential of the Great Lakes-St. Lawrence System—some 8,000,000-hp of which are controlled by the Province of Ontario under the Canadian constitution—that helped bring Hydro into being.

But this is only part of the story. With Niagara Falls so close that metaphorically their roar can be heard and their spray felt throughout southwestern Ontario, it seems only natural that the people of the province should have been interested in their development, particularly since the water-power rights on the Canadian side of the river had not been alienated as on the American side, but were vested in a publicly owned provincial corporation, the Queen Victoria Niagara Falls Park Commission, established in 1885.

Or was this interest as natural as it would appear? Across the Niagara River in the State of New York, which shared the power resources of the noble stream, no great public clamor had arisen for the utilization of the Falls. There, in fact, a skeptical and largely indifferent public had somewhat derisively watched the failure of successive groups of promoters to harness this tremendous power. In any case New York was content to leave the matter to private enterprise. Let capitalists risk their money and take what profits they could make, particularly when the profit potential had so long seemed somewhat less than nil.

So also in the neighboring Canadian province of Quebec. There, as in New York State, the pioneer development of immense hydraulic resources had been left to private enterprise, with little or no public interest manifested one way or the other. And by the late 1890's private capital in Ontario was already engaged in small-scale hydroelectric production and preparing to embark on more ambitious developments at Niagara. Furthermore, nothing in the previous Canadian experience had indicated the virtues of public ownership. While it is true that much of the country's internal development, particularly its canals and railways, had been built with state participation, this had been forced on Canada by necessity rather than by any socialistic leanings. In fact, the one completely government-owned and managed enterprise outside the post office and canals—the Intercolonial Railway—had long been an object of ridicule from one end of Canada to the other because of its general inefficiency, the result of political patronage.

Against such a background it is the more remarkable that "White Coal" and "The People's Power" should have become the rallying cries of a political struggle which had its inception in Western Ontario and kept the province in a state of constant emotional turmoil for the better part of a decade. In part the explanation lies in the situation of southwestern Ontario, that small section of Canada in the triangle between Lake Ontario, Lake Erie and Lake Huron, jutting southward toward the United States. Pelee Island, its most southerly projection, actually touches a more southerly parallel of latitude than the northern boundary of California. From Niagara Falls, its most easterly city, one can travel directly east to Boston; and from Windsor, the most westerly, directly west to Detroit, Chicago and Omaha. Across this broad Ontario isthmus travels 30 per cent of the through railway freight between New England and the American middle west.

The principal features of this region, insignificant in size compared to the rest of Canada, are its climate and its soils. The first permits the growth of trees not found in any other part of the country, the second an agricultural production whose dollar value is greater than that of any other entire province. Bounded on the north by the Shield, and covering an area little more than one hundred by two hundred miles, the Ontario heartland contains the largest single block of arable land in Central or Eastern Canada and has long been a support of the Canadian economy, industrially as well as agriculturally. Blessed with a good climate, rich paleozoic soils deposited by the last glacier; abundant supplies of road and building materials furnished by beds of underlying limestones; natural gas, and even fugitive deposits of petroleum, Western Ontario in fact lacked but one prerequisite for a fully rounded economy—the presence of fossil fuels. Because of the great upthrust of the Shield, itself formed eons before the processes of geology had laid down the carboniferous deposits, coal in Canada is found only on its distant flanks, in the Appalachian formations of Nova Scotia and the Rocky Mountain foothills of Alberta. Either source was uneconomic, even after the coming of the railroads.

In the beginning of European settlement, the lack of coal did not greatly matter. The first settlers in what was to become Upper Canada, and later Ontario, came to the Niagara Peninsula from the Mohawk Valley in New York State prior to and during the American Revolution. They were joined after the 1783 Treaty of Paris by the United Empire Loyalists, colonial Americans who refused to foreswear their allegiance to the British Crown, and during the 1800's by elements of the great westering, land-hungry migration that populated Northern New York, Ontario, Ohio, Michigan and Indiana with no concern for national loyalties. At a still later date, following the close of the Napoleonic Wars in Europe and the War of 1812 in America, began the tide of immigration—German, Irish, Scottish and English —which swelled until by the 1850's all the better farmlands of Ontario had been occupied. Until that time, all the new Canadians, wherever they had come from or in whatever decade, found a land from which they could obtain in abundance everything they required for their needs: mature forests of pine, oak, walnut, beech and lesser trees; streams to provide power for grist and saw mills, and soils which would support sheep, cattle and dairy herds and grow grains, cereals, vegetables and fruit.

Yet by the close of the American Civil War or even a little earlier it was apparent that the greatly favored triangle between the Lakes was actually becoming a "have-not" region. Many factors had combined to bring about this paradox. Closely linked were the rapid growth of population, the distinctly limited area suitable for agriculture, and the gradual industrialization of the region. By the middle 1850's Ontario's farmlands had been largely occupied. During the ensuing decade they were intensively cleared. Industrialization, as in the United States, began in the 1840's and quickly expanded during the 1850's, when more than 1,800 miles of railway were built in the brief span of seven years.

During the period home handicraft shops became factories and small industrial plants were established, using water or wood fuel for steam power. But the clearing of the forest cover reduced both the abundant supplies of fuel wood and the stream flow that had earlier provided dependable water the year round. The familiar phenomenon of spring flood and summer drought appeared, prompting more and more factories to turn of necessity to steam, making ever-increasing demands on the dwindling supplies of wood. It was at this juncture that Ontario had to turn to coal. There were abundant supplies in Pennsylvania, little more than a hundred miles away. It could be bought cheaply and transported at low cost across Lake Erie. But it required payment in American dollars, and American dollars, as they had been from the time of the British Conquest, were in pitifully short supply in Canada.

Until the mid-nineteenth century Canada had received more immigration from the British Isles than did the United States. The influx began at the end of the Napoleonic Wars and was augmented by the Irish famines and depressed economic and social conditions in England and Scotland. Most of the newcomers travelled up the St. Lawrence to Ontario, some to take up land, some to labor on the railroads, and others to find employment as carpenters, mechanics, clerks, bookkeepers and so forth. The trend continued until the brutal nature of the Shield became known. Less than a hundred miles north of Lake Erie, its serrated outcroppings warned of the agricultural wasteland that lay beyond. Some settlers moved there with the lumber industry to homestead but more immigrants moved on to seek their future in the United States, some after pausing briefly, some after a generation spent in Ontario. Encountering the rocky barrenlands of the Shield, Canadian immigration was deflected westward into the adjoining states of Michigan, Indiana, Wisconsin and the Dakotas. For three decades, until the Canadian Pacific Railway had spanned the Shield and the farmlands of the American northwest were finally occupied, the diversion continued. From 1861 to 1901 Canada lost by emigration to the United States some 2,250,000 people—more than she received by immigration from abroad, plus a considerable portion of her natural increment.

Canadian economic historians refer to the period as "The Hungry Decades." It began with the collapse of Jay Cooke's Philadelphia bank in 1873 and continued almost to 1900. By that time the discovery of gold in the African Rand and the Klondike, along with other factors, had improved the world's economy, and Canada's with it. The intervening forty years in Canada had been not entirely years of unrelieved depression. Helped by the protective "National Policy" of Canada's first prime minister, Sir John A. Macdonald, industrialization followed that of the United States, though on a much lesser scale, while the prodigious achievement of a transcontinental railway—the Canadian Pacific—cemented the Canadian Confederation, led to the discovery of the mineral wealth of Northern Ontario, and opened up the Canadian west to settlement. In view of Canada's meagre population and equally meagre capital resources, these were accomplishments of which any people could be proud. Nevertheless, Canadians generally regarded the period as one of unrelieved frustration and stagnation compared with the dynamic progress made by the United States.

In terms of general business conditions, Western Ontario was probably less affected than any other part of Canada by the rigors of the long stagnation. Yet nowhere in the Dominion were there greater feelings of discontent. Other parts of the country could look on the economic stalemate with a feeling of hope deferred; Western Ontario, after its bright beginning, tended to regard its static condition as incurable. The reason for the pessimism lay in its proximity to the United States, of which, but for the international boundary, it was almost part. Scarcely a Western Ontario family had not seen brothers or cousins leave for "the States." None was unaware of the fabulous material progress being made there. And those few who lacked direct connections heard in imagination the beckoning clatter of the foundries and machine-shops of Buffalo, Cleveland and Detroit.

For those Canadians who remained at home, and those in the lake-bound triangle particularly, the United States exerted a mingled attraction and repulsion that grew into a neurosis in which envy and repudiation of the neighboring republic were equally combined. During the depression of 1893 especially, the mood of criticism grew more pronounced. This was the decade during which the United States experienced the Homestead strike and the Haymarket riots, the formation of the great monopolistic trusts, the march on Washington of Coxey's Army of unemployed, and the first muckraking magazine stories about American big business. People in Western Ontario followed these developments, and it is not without significance that both the Knights of Labor and, later, the Populist Movement won adherents in the area. Populism, it may be recalled, had its expression in the People's Party, an American political movement which advocated among other things expansion of the currency, a graduated income tax and the national ownership of all means of public communication and transportation.

Speaking broadly, Canadian sentiment has been more favorable to the Democratic than the Republican Party, largely on the basis of tariff policies. Resentment against the McKinley tariff of 1890 was therefore offset by the election of Grover Cleveland for a second term in 1892 and by his tariff reform of 1894. In 1896 also, during his second administration, came the upturn in the world's economy that ended the long depression. It was signalized in Canada by the defeat of the Conservative Party and the election of a new government headed by the Liberal leader, Wilfrid Laurier. A resurgence of faith in Canada was reflected by the formation of a movement to foster a national culture by a group of intellectuals in Toronto, the provincial capital.

Sir Wilfrid Laurier had yet to receive his knighthood from Queen Victoria or to proclaim that the twentieth century would belong to Canada. Nevertheless, a brightened hope spread through the country as American settlers began moving north across the 49th parallel to the Canadian prairies; manufacturing improved, and the first new railway charters in fifteen years were granted in Western Canada. Western Ontario felt the economic upswing in more jobs and increased production throughout its diversified manufacturing industry. But as prosperity increased, so did its dependence on coal from a foreign source. Psychologically disturbing at worst, the economic hazard of such a situation was dramatically emphasized by an American embargo against the export of coal to Canada following the strike of 75,000 miners in Ohio, Pennsylvania and West Virginia. The strike lasted from July 2 to September 11, 1897, and toward its close the price of coal in Ontario rose to fantastic heights.

It was under these influences that the people of Western Ontario learned, in the depths of the depression, of the successful harnessing of Niagara Falls for the development of electric power. They now began to ask, "If electricity can be carried from the American side of the Falls to Buffalo, why can't it be carried from the Canadian side of the Falls to Hamilton, Brantford, Berlin (the name of which was changed in 1916 to Kitchener) and other cities where it is so badly needed?" No satisfactory answer was forthcoming. There were no large centres of population in Western Ontario and Canadian capitalists were simply not interested in supplying power to such a small and scattered market.

It was in a mood of intransigence, therefore, that a handful of businessmen in Western Ontario set out to secure for their communities the benefits of a great natural resource which was already the people's property—the falling waters of Niagara. The fact that by 1902 close to one hundred Ontario municipalities had organized a total of 126 municipally owned utilities—for the most part waterworks and electric street-lighting services—and had in general experienced little difficulty in their successful management and operation, may have been a source of encouragement to the incipient public-power movement and may have provided it with some useful precedents.

The mood of the time was neither socialistic nor self-consciously nationalistic, although it was undoubtedly tinctured by disapproval of the contemporary monopolistic practices of American capitalism. Its principal ingredient was simply a determination to fill a glaring economic vacuum.

And out of this mood was created Ontario Hydro—the People's Power.

TOOLS OF PROGRESS

2

Had there been no Niagara Falls, no plunging descent of the ponded waters of the Great Lakes after sweeping over the crest of a Silurian escarpment that extends from the vicinity of Albany to the rapids of Sault Ste. Marie, there would in all probability have been no Hydro in Ontario. It was the existence of this great natural source of power, conjoined with the historical influences outlined, that gave the movement for public ownership in Ontario its initial impetus and subsequent direction. The early story of that enterprise is therefore the story of the Falls—from their discovery by the French to their subjugation for human needs toward the close of the last century. To understand the one it is necessary to know something about the other.

On first thought it seems somewhat incongruous that most of the events that were connected with the development of the power potential of the Falls in the two intervening centuries should have taken place on the eastern rather than the western bank of the river; which is to say on what is now American soil rather than on Canadian. Yet this was so natural as to appear almost inevitable. For it was on the east bank at the river's mouth that the French first landed in 1678, and it was there they built their first palisaded fort and trading post, Onguiachra, whence a long-established Iroquois portage route led southward to Lake Erie. Not only was this route more favored by topography than those across the river, it was also more logical and more convenient. So it remained during the years of French and British occupancy, until the transfer of the area to the United States by the Treaty of Paris in 1783. Thereafter, the more rapid growth of settlement along the American frontier plus the more amenable character of the American Falls led to their earlier development.

It is probable that Champlain knew about the great falls of Onguiachra as a legendary phenomenon and that Etienne Brulé was the

10

first white man actually to gaze upon them. Historically, however, the credit for their discovery goes to Father Louis Hennepin, the Recollet missionary-explorer-writer who was a member of LaSalle's first expedition in quest of the Mississippi, and who reached the Falls shortly after sunrise on the morning of December 6, 1678. So impressed was he by the awe-inspiring sight that he wrote these words in his diary under the same date:

> Betwixt the Lakes of Ontario and Erie, there is a vast and prodigious cadence of water which falls down after a surprising and astonishing manner, so much so that the Universe does not afford its parallel.

Yet nearly eighty years passed before another Frenchman, one Chabert Joncaire, Master of the Portage and Director of the French Government's Trade with the West, in 1757 dug a ditch on the eastern bank above the Falls and erected a small mill to provide lumber for strengthening the defences of Fort Niagara, successor to Fort Onguiachra. Two years later the Fort and the adjacent area fell into the hands of Sir William Johnson and his Mohawk Indians. Nothing more is heard about the Falls until the 1790's, when two young surveyors from Connecticut, Augustus and Peter Porter, bound westward with the first expedition to settle the Western Reserve, decided to return there as soon as their surveying duties at the future site of Cleveland would permit. A few years later, in 1805, when the State of New York offered the land along the Niagara River at public auction, the Porter brothers were the highest bidders for the lots which included the American Falls, the rapids above and the beginning of the Gorge below. They also bought river-front lots between the Falls and the village of Buffalo, twenty miles upstream. Title to the land included riparian and water-power rights as well.

With the Porter brothers begins a story of successive efforts, extending over more than three-quarters of a century, to harness the descending waters of Niagara to produce mechanical power and finally to convert their energy into electricity and to solve the problem of its transmission beyond the immediate environs of Niagara Falls. The story is largely one of high hopes and speculative vision, elaborate engineering projects and desperate efforts to find sufficient capital. For more than eighty years these attempts ended in repeated frustrations and disappointments.

The Porter brothers began in 1805 by erecting, on the site of Joncaire's saw-mill, a stone grist-mill and tannery, financed by the profits of their carrying business across the Lewiston-Niagara portage. Disrupted by the War of 1812, the business was rendered obsolete by the opening of the Erie Canal in 1825. The Porters then sought in earnest to turn their ownership of the eastern river bank to profit. Letters patent confirming their power rights were obtained from the State Legislature and eastern capitalists were invited to invest in their development. A brochure issued at the time contains this statement:

> . . . the inadequacy of capital in this part of the country for undertakings of this kind, added to the doubts which until very recently existed in regard to the success of American manufacturing generally, has hitherto prevented improvements which this situation so powerfully invites.

A lack of capital persisted throughout Augustus Porter's life. Although his first proposal envisaged the use of only the fifty-foot drop in the rapids above the American Falls, more convenient sites were available elsewhere and his invitation to eastern capitalists went a-begging. Peter Porter died in 1844, by which time the burgeoning of manufacturing and industry in the East emboldened the aging Augustus, now nearing eighty, to launch a more ambitious scheme: the construction of a huge reservoir on the top of the Gorge down river from the falls, fed by a canal wide and deep enough to furnish water power for the largest manufacturing plants then known.

The Porter heirs were the principal beneficiaries of the patriarch's dream of "a hydraulic canal." In 1853 they sold part of their land and diversion rights to a company formed by a group of Boston and New York capitalists. Five hundred thousand dollars were raised and spent before the first company was succeeded by another, the Niagara Falls Canal Company. By 1860 the second company had run out of money, with work on the intake basin, feeder canal and reservoir still uncompleted. The project was next taken over by Horace and Stephen Allen, two of the original promoters, who took over the old company, reorganized it and prepared to finish the job. They were greatly aided by the new explosive, dynamite, but almost ruined by construction stoppages during the Civil War. Thereafter they completed the work, but not until 1875 were they able to secure their first customer, a flourmill erected by Colonel Gaskill, a veteran of the Civil War.

By this time Day and Allen had spent most of their own money and mortgaged the land and water rights owned by the Company. Early in 1877 the creditors closed down and the engineering work on which three groups of promoters had spent upwards of $1,500,000 was sold on May 1 at public auction by the Sheriff of Niagara County for the sum of $71,000. The buyer was Joseph Schoellkopf, a Buffalo manufacturer who had arrived in America as an immigrant with $800 in 1844.

Public interest in Father Hennepin's prodigious cadence, however, was by no means confined to industrial enterprise. By the time of the Schoellkopf purchase the Falls had become the greatest scenic attraction in the United States. As early as 1826, when the Erie Canal brought the first sightseers from the east to view the stupendous spectacle, Augustus Porter had recognized the commercial possibilities of the tourist trade. A bridge was built joining Goat Island and the mainland, steps were erected and paths laid out leading to lookouts commanding views of the American Falls and the greater Horseshoe Falls that belonged to Canada.

The newly built railroads of the 1840's brought increased hordes of trippers. By the close of the decade the town on the American side of the river was known as "The Honeymoon Capital" and there were plans for a suspension bridge to join the American and Canadian shores. But while the Falls had become a magnet of scenic attraction, they had also become a centre of intolerable huckstering. Extortion was practiced alike by the hackmen who infested the railway depots and the concessionaires who levied a toll for admission to commanding sites—where these were not blocked by the unsightly mills, breweries and distilleries that had sprung up along the river bank.

By 1869, so outrageous was the defilement of the beauty of Niagara that a group of public-spirited Americans, including such notables as Frederick E. Church, the artist, landscape architect Frederick Law Olmsted, and the architect, H. H. Richardson, were moved to vigorous protest. The interest was enlisted of a former lieutenant-governor of New York State, the Hon. William Dorsheimer, and efforts were made at Albany to have the State Assembly and Senate enact legislation declaring the area above the Falls a state reservation for scenic preservation and public enjoyment. Despite the eminence of the sponsors and a growing body of public opinion, the attempt failed. Year after year bills seeking to preserve the beauty of Niagara were introduced in the State Legislature. Year after year they were defeated with monotonous regularity by more powerful interests.

It remained for Lord Dufferin, the Governor General of Canada, to break the deadlock. When Frederick Church appealed to him on artistic grounds, the Irish earl addressed the Ontario Society of Artists at Toronto in 1878 and advocated the creation of "a small international public park" at Niagara Falls. The speech was widely publicized in the United States and was so favorably received that Governor Robinson of New York (with whom Lord Dufferin had been in correspondence) in 1879 sent a message to the Legislature advocating the purchase by the State of land along the Niagara River and Falls to be used as a State park and reservation. A year later the Dominion

Government passed an "Act Respecting Niagara Falls and the Adjacent Territory" which gave the Federal Minister of Works the right to acquire lands for park purposes, but no action was taken. In 1883, however, the New York Legislature passed an act establishing the New York State Reservation at Niagara Falls, and two years later Lord Dufferin's recommendations were finally realized when the Ontario legislature created Queen Victoria Niagara Falls Park.

Joseph Schoellkopf was not immediately affected by the benign international conspiracy to preserve the beauty of Niagara Falls. That would come later, after the physical establishment of the New York State Reservation had restricted the area available for industrial use. Meanwhile, Schoellkopf's problem was the same as that which had faced his predecessors: raising the capital needed to continue the work where Day and Allen had left off. This was accomplished by the formation of a third company, the Niagara Falls Hydraulic and Manufacturing Company, and the sale of sufficient stock to proceed with the great project. More customers were also sought, and so favorable had the manufacturing outlook become after the long depression of the 1870's, that by 1882 Schoellkopf and his associates had built a small powerhouse, Station One, and were supplying seven mills with water power.

All these mills were run by direct water-power operating turbines which in turn drove machinery by means of shafts, belts and pulleys. And then in the closing weeks of 1881 came an epochal event: the installation of an electric generator by the Brush Electric Light Company. Up to this time electricity had few commercial uses except to operate telegraphs and telephones. The principle of the direct current motor had been discovered only eight years before. Less than two years had elapsed since Jablochkov had demonstrated his electric candle, Charles Brush his open arc-light and Edison and Swan the incandescent lamp. The genius of Menlo Park had yet to install his first central power station in New York City and Nikola Tesla to embark on the brilliant series of inventions which included the alternating current motor, transformer, and high-voltage transmission.

The Brush generator was one of the first hydro-electric installations anywhere in the world. Driven by Niagara power, it produced enough current to operate sixteen arc-lights. Electricity had come to Niagara but, transmitted by direct current, its use was still limited to a small radius.

Another problem soon arose: that of factory sites for the use of the hydraulic power available. The Schoellkopf development got its water from a surface canal, as projected by Augustus Porter. It ran overland through the village of Niagara Falls and furnished power only on the banks of the terminal basin. Expansion at that location was not limited by the projected State Reservation, the land for which had been acquired from the heirs of Augustus Porter for $1,439,423.

It was at this point that Thomas Evershed, a division engineer on the Erie Canal, in a letter published in the *Lockport Union*, proposed a greater use of Niagara's hydraulic potential than any yet conceived. For the period, the project was truly grandiose. It involved the building of a huge subterranean tunnel, to run for two miles and a half beneath the town of Niagara Falls, New York, and serve as a common tailrace for thirty-eight turbines sunk in vertical shafts. Their combined output, some 119,000-hp, would be sufficient to turn the shafts and pulleys of more than two hundred factories located on the ground above.

The Evershed plan fired the imagination of a group of businessmen, who shortly afterward formed the Niagara River Hydraulic Tunnel, Power and Sewer Company, with Charles B. Gaskill as its President. A charter was granted by the State Legislature in 1886. Once again history repeated itself. A search for capital took the promoters first to London, where all attempts to interest British investors failed completely. In Wall Street they fared better. In 1889 the Company was reorganized and renamed the Niagara Falls Power Company. An affiliate, the Cataract Construction Company, was formed to act as the financial agent, with Dean Adams of the New York banking firm of Winslow, Lanier and Company as President, and the distinguished American engineer Dr. Coleman Sellers as Engineering Consultant.

When completed, the installation envisaged by Sellers would produce unprecedented amounts of power. But would it find a market? Not, it would seem, in the town of Niagara Falls, New York, with a population of only 3,500 people. But twenty miles away was the city of Buffalo, a major shipping and industrial city with a population of a quarter of a million. Buffalo could use all the power that Evershed's thirty-eight turbines could produce—if only there was some way to get it there. In 1890 no way was known in America.

With the hope of stimulating discovery, public-spirited citizens throughout the country offered a prize of $100,000 for a solution to the baffling problem of long-distance transmission. Those directly interested in the Evershed scheme also tried to find an answer to the unsolved riddle by inviting opinions from scientists and engineers the world over. No solution was forthcoming, but out of the preliminary search came three important and far-reaching decisions: to retain the Evershed plan of a subterranean discharge tunnel; to develop power at a single station; and to generate that power in the form of electricity.

In the fall of 1890 President Adams and his consulting engineer, Dr. Sellers, sailed for France. During the following months they inspected every important hydraulic and electrical installation in Europe. In Geneva they saw water power being distributed by pressure through a pipe system but the length of transmission was less than a mile. In Berlin they inspected a 1,000-hp direct current generator, the world's largest, but its power was transmitted less than half a mile. In Budapest they examined a steam-driven generator which produced alternating current, but only for arc-lighting within a limited radius.

Baffled, the two Americans resolved upon a research as magnificent as the Niagara project itself. In London a consulting board called the International Niagara Commission was established in Brown's Hotel where a plaque now commemorates its meetings. As Chairman, Adams appointed no less a personage than the distinguished physicist Sir William Thomson, later Lord Kelvin. Other appointments included some of the ablest physicists and engineers of the time—among them Professor W. C. Unwin as Secretary, Lieutenant-Colonel Theodore Turrentini of Switzerland, Professor Elie Mascart of the College de France and Dr. Coleman Sellers of the U.S.A.

Through the Commission the Niagara Falls Power Company offered prizes for a solution of its transmission problems, including awards for the best electrical plan, the best hydraulic plan and the best combined hydraulic and electrical distributing system.

Of the many plans submitted none was judged worthy of the Grand Prize. Smaller prizes were awarded for a variety of plans, hydraulic, pneumatic and electrical. Only one of the prizewinners recommended a system based on multi-phase alternating current. But when the Commission report was published, it favored some adaptation of the alternating current systems then being built experimentally in Europe.

In 1884 Gaulard and Gibbs had successfully demonstrated a 25-mile test line from Lanzo to Turin. In 1891 another experimental line, built for the Frankfort Exhibition, transmitted 300-hp of three-phase alternating current at 30,000 volts for a distance of more than one hundred miles. Despite these achievements, the members of the Commission were divided: the younger men, who had nothing to lose but their credibility, favoring alternating current; the older ones, with vested interest in the established systems, direct current. Among the advocates of alternating current were the engineers who had built the Lauffen-Frankfort line: Ferranti, who was then setting up an alternating current system in London; Nicola Tesla, the inventor, and George Westinghouse, who with associates had secured rights to Tesla's patents. Sir William Thomson stood for direct current, as did Thomas A. Edison. Remembering that as late as 1889 Westinghouse had recommended that Niagara power be transmitted to Buffalo pneumatically, the Commission did not find it easy to decide between these conflicting counsels.

Two seemingly dissociated events in 1893 helped tip the scales finally in favor of alternating current. The first was the tremendous impression created by the electrical illumination at the World's Columbian Exhibition at Chicago, installed by Westinghouse and using alternating current. The second was the firm contract for a large block of electrical power, signed by Charles Martin Hall, the inventor of the electrolytic process for the manufacture of aluminum, on behalf of his then-struggling enterprise, the Pittsburg Reduction Company. The contract provided for the delivery of 1,500-hp by February 1, 1895, with an additional 5,000 optional by August 6 of that year.

With Hall's signature on a contract, it now became possible to secure financing for one of the boldest engineering adventures ever undertaken: the building of a transmission line to Buffalo, with all its ancillary facilities and installations.

On October 28, 1895, Adams placed contracts with the Westinghouse Company and with General Electric for dynamos, transformers and the transmission line to Buffalo, and with Swiss firms for Fourneyron-type hydraulic turbines. The largest a.c. generators then in use were 150-hp; yet the Westinghouse Company contracted to build three generators of 5,000-hp each. Equally undaunted, the General Electric Company undertook to build an 11,000-volt overhead transmission line to Buffalo and supply transformers of 1,250-hp capacity at a time when the largest transformer in use was one of only 10-hp.

Although power was delivered to the Pittsburg Reduction Company in August 1895, it was not until November 1896 that all contracts were completed. The system was then tested and put into actual operation on the night of November 15, when Mayor Jewett of Buffalo threw the switch that enabled Niagara Power to energize the local transmission system. So unprecedented was the nature of the event, and so uncertain was its success, that the ceremony was conducted quietly and almost without public notice. Only the mayor, a few officials and engineers of the Niagara Falls Power Company, and a few invited guests were present. Only one newspaper, the *Buffalo Commercial Advertiser*, reported the event.

At a commemorative banquet held in Buffalo in January 1897, the achievement was acclaimed as a triumph of the collaborative scientific and engineering brains of America and Europe. The acclaim was far from extravagant. A little more than a century before, James Watt's reciprocating engine had ushered in the first industrial revolution. The surge of power over the twenty-mile transmission circuit from Niagara marked the dawn of a second, more tremendous in its consequences than the first. Without the long-distance transmission of electricity neither the modern mass assembly-line nor its offspring, automation, would have been possible.

Not only was the Niagara-Buffalo System the world's largest by far at the time, but its versatility was then unique. Part of the two-phase alternating current generated at the Falls was converted into three-phase for high-tension transmission to Buffalo, where it was again converted to supply Buffalo's existing direct current distributing system. Another part was converted into direct current at the Falls for the aluminum pot-lines, to power the street railway, and to supply other customers with arc or incandescent lighting installations. In addition, unforeseen market demand was provided by the electro-metallurgical and electro-chemical plants that were attracted to Niagara by the promise of cheap power.

With the assurance of a market, the Schoellkopf plant now moved to expand. A new Station 2 was built to utilize water and riparian rights confirmed by the New York State Legislature in 1895. Built in the Niagara Gorge at the foot of the escarpment, it utilized the full head of 210 feet to operate turbines which activated two or more generators by means of horizontal shafts. The output of the station's 18 Westinghouse generators—18,000-hp of direct current at 300 volts—was taken by the Pittsburg Reduction Company, so rapid had been its growth. Station 2 also supplied the Niagara Gorge Railroad and other users of direct current.

Niagara Falls, New York, was now booming. During the nineties its population jumped from 3,300 to 19,500; during the next decade it grew at double the rate of the national average.

The successful transmission of electrical energy from Niagara Falls to Buffalo in November 1896 made it possible to foresee a time when even the most remote and isolated hydraulic resources could be made to serve man's industrial and social needs. This development did not go unnoticed in the nearby Province of Ontario. There, thanks to the aesthetic sensitivity of a cultured Irishman, Lord Dufferin, similar developments were soon to be promoted under the banner of public ownership.

3

As recently as 1870 not a single horsepower of electricity was produced in Canada or the United States for lighting or manufacturing purposes. Yet by 1880 no fewer than 5,453 production units were in use in the two countries, of which 5,110 were in the United States and 343 (approximately 6.8 per cent) in Canada. It is therefore evident that Canada, with a population then of 4.83 millions as against 62.9 millions in the United States (some 7.1 per cent of the total of the two countries) was not far behind her great neighbor in making use of the new form of energy as soon as its practical application became possible.

Canadian interest in electricity was first prompted by the country's great distances and by the slowness of communications, just as at a later date interest in the use of electricity for industrial and manufacturing purposes would be prompted by the uneconomic distribution of fossil fuels. As early as 1846, only two and a half years after Morse sent his first electric telegraph message from Washington to Baltimore, a Canadian company, the Toronto, Hamilton, Niagara and St. Catharines Electro-Magnetic Telegraph Company, was transmitting messages between those points. A few months later the lines were extended to Buffalo, establishing connections with points in the United States. By the end of 1847, the Montreal Telegraph Company had strung its wires from Toronto on the west to Quebec on the east, a combined distance of some 525 miles, and a third Canadian company, the British North American Electric Association, was preparing to build from Quebec City to the Atlantic coast.

Telegraphs were at that time still operated by batteries and it was not until 1870, when Gramme patented his invention of the dynamo or generator, that electricity passed from laboratory research and experiment to commercial utility. Three years later at the Vienna Exhibition it was shown that a Gramme-type dynamo acted as a motor

19

when set in rotation by the current from a similar machine. The Vienna demonstration was repeated at the Philadelphia Centennial Exhibition in 1876, and that same year Paul Jablochkov, a Russian officer living in Paris, invented his "electric candle"—an open arc-lamp consisting of two rods of carbon which produced a brilliant light. Within two years Charles F. Brush of Cleveland, Ohio, produced a more effective type of dynamo, and became the pioneer of arc-lighting in the United States, installing lamps invented by himself.

The inventions of Jablochkov and Brush mark the beginning of commercial arc-lighting. With them, laboratory experimentation entered the industrial phase: the production of electrical energy for commercial purposes. Within a short time more than a dozen arc-lighting systems were in use, the foremost being those of Brush and Thomson-Houston.

The rapidity with which the use of the new utility was promoted in Canada is astonishing. The year 1878 saw the formation of the Edison Electric Company in the United States, as well as the introduction of the first Canadian company in this sphere, the American Electric and Illuminating Company, and the building of a tiny generating plant in the retail business district of Montreal, Canada's commercial metropolis. That same year the Jesuit Fathers on Bleury Street in the same city are said to have received from France a Jablochkov "candle" which was lighted by telegraph batteries. Other authorities claim that the first electric lighting was supplied by two arc-lights installed in Haymarket Square while others give precedence to a lamp at the entrance to St. Lawrence Hall, a famous mid-Victorian hostelry on St. James Street, placing the date as 1879.

Whether the date was '78 or '79 matters little; the enterprise of Canadian capitalists with respect to the new source of illumination is self-evident. This enterprise was given impetus by a provision in Canadian law which required that goods made under Canadian patents be manufactured in the country. As a consequence Thomas A. Edison, shortly after securing his American patents, came to Montreal to produce and install the first Canadian incandescent light in a paper-box and wire factory belonging to Messrs. Major and Gibbs on Craig Street. Two years later, in 1881, the Ottawa Electric Light Company was formed in the nation's capital where three dynamos driven by a 16-hp water-wheel lit arc-lights in a few mills.

Hamilton, at the head of Lake Ontario, was the first Canadian city to install street lighting. In the same year, 1883, Edison returned to Canada to install his system in the plant of the Canada Cotton Company at Cornwall on the St. Lawrence River where it provided lighting for a new weave shed.

Toronto, the provincial capital of Ontario, did not enter the electrical story until this time, when John Wright, an English millwright, installed arc-lights on three of the principal thoroughfares: Yonge Street, from the waterfront to Wilton Avenue, and King and Queen Streets, from Church to York Streets. Wright had come to Toronto in 1870 and in 1876 had visited the Centennial Exhibition in Philadelphia where he attended lectures on electricity conducted by the Philadelphia Technical Institute three nights a week. Two of his teachers were Elihu Thomson and Edwin J. Houston, then working on their self-adjusting dynamo, and such was Wright's skill that he entered their employ and worked for them for several years. On his return to Toronto he was given, in recognition of his services, two dynamos, which became the city's first generating plant. Operated by surplus steam from the plant of the *Toronto World*, a morning newspaper, these were used to light the arc-lamps which Wright had installed on the streets. Since the lamps were not needed during the day time, he conceived the idea of building one-quarter horsepower motors as labor-saving devices for the home. One was demonstrated in a diminutive shop next door to the newspaper office, to turn a coffee grinder. Curiously enough, his enterprise was greeted with violent opposition as an invention of the devil, and one Congregational minister delivered a sermon in which he denounced the electric motor as an instrument of evil since it would release girls from honest toil to wander about the streets and fall prey to the wiles of Satan.

Undismayed, Wright went on to build and demonstrate one of the world's first electric railways in 1883, a one-mile line which ran from Strachan Avenue to the Toronto Exhibition Grounds on the western outskirts of the city. There is much confusion in existing historical writing on the subject as to the actual date on which the first electric railway was operated, most authorities giving the year as 1885, when such a railway was put in operation between Baltimore and Hampden, Maryland. However, a reference to the *Toronto Daily Mail* for September 12, 14, 17 and 21, 1883, clearly establishes the fact that a line was built and operated that year, if only for a short period.

The records of the 1880's also contain such advances as the introduction of incandescent lamps for street lighting in Cornwall, Vancouver, and Victoria; the installation at Pembroke, Ontario, a lumbering centre on the Ottawa River, of an Edison generator; the change from gas to incandescent lamps in the Parliament Buildings in Ottawa;

and the introduction of commercial electrical service in Montreal, where gas had been used since 1837. Most of these events took place in 1884 and the next year an Ottawa promoter, Thomas Ahearn, signed a contract to light the streets of the capital city by arc-lamps. A clause provided that service would not be supplied on moonlit nights. Later in the decade a 100-hp generator was installed at a dam on the Credit River, a small stream flowing into Lake Ontario between Hamilton and Toronto, and power conducted by copper wire to a paper mill two miles distant. Regarded in local engineering circles as an important contribution to the new technology, the initial hydro-electric plant was duplicated not long after by two others, one on the Credit River and one on the Pine. In the context of the wider aspects of electrical development, the record of progress throughout the Dominion is of Canadian historical interest only; nevertheless, it was sufficient to enable A. J. Lawson, one of Canada's pioneer electrical engineers, to state in 1890, in an address to the Canadian Society of Civil Engineers, that there was hardly a village in Ontario of over 3,000 inhabitants that did not have an electric-light station of some kind in operation and that few of the important towns in the other provinces were without electric lighting. In use in Canada at the time, according to the same source, were 13,530 arc-lights and 70,765 incandescent lamps.

Overshadowing any physical progress of the 1880's, however, was the crucially significant event referred to in the previous chapter—the creation of the Queen Victoria Niagara Falls Park by the Ontario Legislature in 1885.

By the terms of the British North America Act of 1867 establishing the Dominion of Canada, the Federal Government was given control over navigable waters, but ownership of the beds of lakes and streams was vested in the provinces. The consequent ambiguity gave rise to one of the most celebrated legal and legislative battles in Canadian history as, at an earlier date, an identical ambiguity in the American Constitution had given rise to litigation between New York and New Jersey, resulting in Chief Justice John Marshall's famous decision placing U.S. navigable waters under the control of the Federal Government.

At stake in Ontario was the right of the Province to regulate the lumber industry, then one of the most important factors in the Canadian economy and, along with it, the ownership of hydraulic power resources. Brought to a successful conclusion by Oliver Mowat, Premier of Ontario from 1872 to 1896, the Canadian controversy was concluded by a decision of the British Privy Council which gave to

Ontario the rights demanded and which included the lion's share of the flow of the Niagara River and hence its great hydro-electric potential.

Mowat, known as the "Defender of Provincial Rights," in 1881 introduced the "Rivers and Streams Bill," designed to protect the public rights of the waterways of the province, one year after the Provincial Government had passed "An Act Respecting Niagara Falls and the Adjacent Territory" which gave the Minister of Public Works of the Dominion of Canada the right to acquire lands for park purposes. Disallowed three times by the Dominion Government, the Rivers and Streams Bill was each time re-passed by the Provincial Legislature, the conflict in jurisdiction being finally settled in Ontario's favor by decision of the Privy Council in 1884. During the succeeding year, 1885, the Ontario Legislature passed "An Act for the Preservation of the Natural Scenery about Niagara Falls." Known as the Niagara Falls Park Act, and similar in its purpose to the New York legislation of 1883 creating the New York State Reservation, the Ontario enactment gave the Lieutenant-Governor power to name a board of three commissioners to carry out the provisions of the Act. Within a month Sir Casimir Stanislaus Gzowski, John Grant Macdonald, and John Woodburn Langmuir were appointed, with Colonel Gzowski to act as Chairman. A fourth commissioner, John A. Orchard, was added in 1887 when the Commission was made a provincial public corporation and the name Queen Victoria Niagara Falls Park was adopted. Authority was granted to issue $525,000 worth of forty-year debentures. The Commission promptly acquired the lands bordering the river, abolished the objectionable features of tourist exploitation, prohibited gambling and the sale of liquor and ended the pestering of hucksters and cabbies within the Park.

Thus, about the same time that the Niagara Falls Power Company began the development of the American Falls under the rights descended from the Porter brothers, public ownership and control of Ontario's much larger share of Niagara's power was firmly established; not, however, as a hydro-electric resource, but as a scenic monument whose beauty was to be conserved for the enjoyment of future generations.

This, the record shows, was the Commission's first concern. But the upkeep of the Park was costly. Income from concessions soon proved inadequate, and the Provincial Legislature was loath to burden Ontario's taxpayers merely to delight the eyes of nature-lovers and honeymooners at the Falls. Nor did the legislature at this time share the enthusiasm for public ownership displayed by William Dean Howells, the American novelist, who paid warm tribute to the transformation achieved by the New York Park Commissioners:

> I never greatly objected to the paper mills on Goat Island. They were impertinent to the scenery, of course, but they were picturesque, with their low-lying, weather-worn masses in the shelter of the forest trees, beside the brawling waters. But nearly every other assertion of private rights in the landscape was an outrage to it. I will not even try to recall the stupid and squalid contrivances which defaced it at every point, and extorted coin from the insulted traveller at every turn. They are all gone now, and in the keeping of the State the whole redeemed and disenthralled vicinity of Niagara is an object lesson in what public ownership, wherever it comes, does for beauty.

Thus early do we hear sounded the theme of Beauty and the Beast of private profit with which Ontario legislators, for the next generation, would become all too familiar. But could not public and private interest co-exist in Niagara? Could not the Commission safely accept the good annual rentals which American entrepreneurs were soon to be prepared to pay for long-term leases of water rights at the Canadian Falls? The Commissioners were strongly tempted.

The first entrepreneurial interest in Canadian Niagara power occurred in 1887, when an American syndicate formed the Canadian Power Company (later the Ontario Power Company of Niagara Falls) and applied for and received a charter from the Dominion Government to develop water power from a canal to be built from the Welland River or Chippawa Creek to the Niagara Gorge. The company encountered financial difficulties and did nothing for the next five years.

Again in 1888 the Park Commission was approached by a group of American capitalists who offered to pay a yearly rental of $25,000 for ten years, and after that an increase of $1,000 a year until the annual rental of $35,000 should be reached. But while negotiations were still in progress, the American promoters withdrew their offer because of differences within the group. They were quickly succeeded by a group of English financiers "with unlimited capital" associated with the electrical transmission pioneer, S. Z. de Ferranti. The new group made a deposit of $10,000, but by March, 1891, when their deposit became subject to forfeit, they had done nothing. The option was then extended until March 1, 1892, when it was salvaged by Edward Dean Adams and the group of American capitalists then engaged in building the Evershed tunnel across the river. A month later the Commission signed a contract with this group, incorporated as the

Canadian Niagara Power Company, a subsidiary of the Niagara Falls Power Company.

That same year, the Queen Victoria Niagara Falls Park Commission, in need of revenue and anxious to provide transportation for sightseers, awarded a franchise to the Niagara Falls Park and River Railway. The company undertook to build and operate an electric railway from Niagara Falls to Queenston and to pay a yearly rental of $10,000, and installed a generating plant to produce 2,100-hp. With Commission approval, the railway company later sold its surplus hydraulic power to the Canadian Niagara Power Company.

The twenty-year contract with the Canadian Niagara Power Company conveyed exclusive rights to the use of Niagara water on the Canadian side for twenty years, with an option to renew for four further twenty-year periods, constituting a hundred-year monopoly. The company paid two years' rent—$50,000—in advance and agreed to start construction in May 1897 and to complete by November 1898 a hydraulic installation to develop 25,000 electrical horsepower. It also agreed to supply Canada with current "to the extent of any quantity not less than one-half the quantity generated."

May 1897 arrived and still construction had not begun, although the company had paid the Commission $100,000 in rent. Pleading difficulty in solving the long-distance transmission problems, although the parent company's Buffalo transmission line was then in operation, the Canadian subsidiary obtained from the High Court of Justice for Ontario a twelve-month extension, making November 1899 the agreed date for the installation and delivery of 10,000-hp.

As the new deadline approached and the Canadian Niagara Power Company was clearly not going to be able to fulfill the terms of its contract, the Ontario Legislature passed an act giving the Park Commissioners authority to bring about the complete surrender or abandonment of the monopoly rights which the company had enjoyed under its lapsed contract.

Once the monopoly clause was abrogated, the Commissioners were prepared to make concessions. A new contract was written, embodying a franchise good for fifty years, with an option to renew for three subsequent periods of twenty years each; rent and other conditions to be adjusted with each renewal. It was stipulated that 10,000-hp of electricity was to be available by July 1903. Again the company failed to meet its deadline, but as it now proposed to install initial machinery to produce 50,000-hp rather than 20,000, and to complete the whole works so as to provide water connections for the full ultimate amount, it was granted a further extension to January 1, 1905.

The granting of monopoly rights to the Canadian Niagara Power Company in 1892 had been strongly criticized in the press and from the public platform in Ontario: the contract signed in 1897 brought protests from another quarter—the moribund Ontario Power Company, chartered by the Dominion Government in 1887. Fearing that its charter rights might be extinguished, the company moved to renew its charter. It was successful, and in 1900 secured an agreement with the Park Commission to use water from the Welland River to develop 180,000-hp, the license to be for fifty years. A contract was signed, but two years later it was changed to permit the use of water from the Niagara River, despite the protests of the competing company.

At this juncture, when actual construction on the elaborate engineering works foretold the prospect of eventual production, the Canadian Niagara and the Ontario Power Companies were joined by a Canadian syndicate, the first such to show an active interest in the development of Niagara Power. The group was made up of William Mackenzie, a celebrated Canadian railroad builder; Henry M. Pellatt, a Toronto investment broker and promoter; and Frederic Nicholls. The latter, the youngest of the group, had come to Canada in the early eighties, secured a position with the newly-formed Canadian Manufacturers Association, and in that position had taken an early interest in electrical development. He was one of the founders of the Toronto Electric Light Company and of the Electrical Manufacturing Company, a firm closely associated with the General Electric Company in the United States and which later became the Canadian General Electric Company.

Mackenzie, Pellatt and Nicholls were all residents of Toronto and promoters of formidable resources and ability. They controlled numerous enterprises, including the Toronto Street Railway and the Toronto Electric Light Company, then the largest users of power in Ontario. In January 1903 the syndicate they headed was granted a franchise to develop 125,000-hp on terms similar to those granted the other two licensees.

With these projects to generate close to 400,000-hp now progressing beyond the planning stage, the approaching era of full-scale exploitation of Niagara's hydraulic resources could be clearly foreseen. As scientists solved the basic technological problems, however, the political question of public or private ownership assumed ever-increasing importance—and in 1903 the odds seemed strongly in favor of the private interests.

WHY NOT PUBLIC OWNERSHIP?

4

At the turn of the century Ontario industry, most of it concentrated in the narrow agricultural belt north of Lake Erie, was experiencing a power shortage. Wood, which had once fueled the boilers of its foundries and factories, was no longer abundant. Coal, which had to be mined and shipped hundreds of miles from Nova Scotia or Alberta, or imported from the United States, was becoming increasingly expensive. Water power, immense and inexhaustible, was near at hand. But the water wheel, except at a few strategic sites, was not an efficient power converter. Without the turbine, generator and high-voltage transmission line, Ontario's splendid hydraulic resources would continue to pour almost unused from the myriad collection basins of the Pre-Cambrian Shield to the cliffs of Niagara and thence down the St. Lawrence River to the sea.

Hence when Buffalo's mayor, on the night of November 15, 1896, turned the switch that brought cheap electric power from the Falls to Buffalo, Ontario's cities and industries did not share in the rejoicing. They foresaw defeat in their unequal efforts to compete with low-cost American producers. Already the Canadian economy was being seriously drained by the capital exports required to purchase coal from the mines of Pennsylvania. And in 1902, when the Pennsylvania miners struck, and the privately owned railways refused to move from the pit heads what little coal was available, the situation for Ontario's manufacturers became disastrous. Coal which before the strike had cost the city of Toronto $3.50 a ton was now hard to get at two or three times that price.

For a while, Toronto imported coal from Wales at $10 a ton, knowing this to be the only stopgap. Without a firm source of power, either coal or hydro-electric, the situation was bound to remain precarious. Without power such as Buffalo was getting from the new installations at Niagara Falls, Toronto's advantages as a combined railway terminal

27

and lake port would not be enough. Nor could the smaller cities of Ontario—Hamilton, Brantford, Guelph, Galt, Berlin—compete without generators and transmission lines to utilize the hydraulic power sites on the Grand River, the longest stream in the southwestern part of the province.

During a half-century of slowly accelerating industrialization the landscape of Ontario had become dotted with saw-mills, planing mills, hoop and stave factories, furniture factories, flour- and grist-mills, tanneries, foundries, machine shops, brick and tile works, textile mills. Most of these plants were steam powered, using some 4,000 boilers and about as many stationary engines. Cheap hydro-electric power, coupled with the flexibility of the electric motor, would stimulate all these industries.

But the power had to be cheap. Moreover, it had to be readily available wherever it was needed, freed from the limits imposed by the self-interest and caution of the capitalistic concerns which controlled most of Canada's electrical power development.

Without exception, the attitude of the private electrical interests was both arrogant and avaricious, of a piece with Cornelius Vanderbilt's famous dictum of a generation earlier, "the public be damned." Having the foresight to pioneer electrification, and the modest capital required to implement their initiative, those interests had secured a monopoly position and, not unnaturally, like their mercantilist forbears, were determined to exploit that position to the utmost in terms of profits. Viewed in retrospect, their "customer relations," or, better, lack of them, are seen to have been nothing less than suicidal. So bad were they, in fact, that the first protests against them came not from the small domestic consumer but from Ontario's manufacturing industry itself.

One of the earliest critics of both the alienation policy of the Queen Victoria Niagara Falls Park Commission and of the public attitude of the Toronto Electric Light Company was W. W. Maclean, the intransigent publisher of the *Toronto World*, and a long-time member of the Dominion House of Commons. Another was Dr. Beatty Nesbitt, an outspoken member of the Ontario Legislature, who returned from a privately financed visit to hydro-electric installations in Europe and the United States to proclaim his conviction that a publicly owned power system would revolutionize the industrial economy of Canada.

By the year 1899, advocacy of public ownership was heard even in the lounge of the conservative Toronto Board of Trade. The protagonists were two well-known Toronto citizens, P. W. Ellis and W. K. McNaught. Both were manufacturing jewellers and, like Nesbitt, both had been to Europe. In Switzerland, then the world's most advanced country in industrial electrification, they had seen what cheap hydroelectric power had done for their Swiss suppliers and correspondents. On their return, they urged that the City of Toronto consider constructing its own generating station at the Falls, its own transmission lines to the city, and even its own distributing system. They were not interested in public ownership *per se*; but simply obtaining power at rates lower than those charged by the local electric company.

That same year the Toronto City Council applied to the Ontario Government for authority to establish a municipally owned public utility to supply the city with electricity for light and power. Promptly the private utility lobby threw back the attack for a loss. Not only did the Legislature reject the application; it enacted the Conmee Clause of the Ontario Municipal Act, the effect of which was to prevent any municipality from competing with a private company engaged in public service until it had bought out the existing privately owned system at a price fixed by arbitration.

The small band of public-ownership advocates refused to be silenced. Almost immediately, agitation was started for the repeal of the Act, and this pressure continued for years afterward. Leading the publicownership faction on the Toronto City Council was Alderman F. S. Spence, described by a contemporary as "a light, angular man who leaned forward as he walked and hurried to keep himself from falling." A Methodist in religion, a Liberal in politics, and an ardent Prohibitionist, Alderman Spence was an effective platform speaker with a considerable following. Among his principal supporters on the Council were Aldermen Urquhart and Ward. In 1900, after the Legislature had rejected for the second time the city's petition for authorization to enter the electric light and power business, Alderman Urquhart in the City Council proposed a resolution, seconded by Alderman Ward, requesting Toronto's Board of Control to inquire and negotiate concerning: (1) the price per horsepower at which electricity could be delivered at the limits of the city of Toronto from works at Niagara, (2) the cost of constructing the necessary works for receiving and transmitting electrical energy within the limits of the City of Toronto, and (3) the cost of transmitting electricity to users within the limits of the City of Toronto.

Two months later, the Toronto Board of Trade took similar action, when a special committee was appointed to study the entire question and report. Its Chairman was W. E. H. Massey, a leading agricultural implement manufacturer and President of the Massey-Harris Company; the other members were Elias Rogers, a Toronto coal merchant; William Stone, head of a lithographic plant, and A. E. Kemp, a sheet metal manufacturer. The Committee's report, rendered three months later, was by no means a public-ownership manifesto. Mildly, the Committee registered its hope for cheaper power, to be obtained by buying current from one of the Niagara generating stations. It asked: should the Toronto Electric Light Company, which had signified its intention of bringing power from Niagara Falls, control this development, or should the City take it over?

Similar questions were being asked by businessmen and municipal officials in other parts of Canada. In 1901 representatives of Canadian municipalities convened in Toronto to discuss ways and means of accelerating the distribution of electric power and controlling its price. Organizing themselves as the Union of Canadian Manufacturers, they elected O. A. Howland, Mayor of Toronto, as President and W. D. Lighthall as Honorary Secretary. Almost immediately the Union's influence became apparent in the cloakrooms of the Parliament at Ottawa, where private-utility lobbyists found themselves for the first time confronted with a semblance of organized opposition. Both in Ottawa and in the provincial capitals the efforts of private companies to control and exploit electrical technology were increasingly challenged by the growing body of opinion in favor of public ownership.

In Toronto the ratepayers voted in 1900 to take over the privately owned Toronto gas works. The issue was settled by a compromise by which the company agreed to permit the City to buy stock at par and seat its mayor on the Board of Directors of the Company. Another private utility, the Toronto Street Railway, was less accommodating. Its running battle with the City Council came to a climax when it became known that William Mackenzie, who controlled the street railway and the Toronto Electric Light Company, was planning to install a generating plant at Niagara Falls to supply power for both the railway and for lighting service.

Before the public-ownership advocates could organize their opposition, Mackenzie and his associates on the directorate of the Toronto Electric Light Company started negotiations with the Niagara Falls Power Company to supply the necessary power. To construct the transmission line, the promoters organized another affiliated company, the Toronto and Niagara Power Company, incorporated by an Act of the Dominion Parliament. When negotiations with the Niagara Falls Power Company broke down, the Syndicate, as the Mackenzie group was called, applied to the Queen Victoria Niagara Falls Commission for a franchise to build a 100,000-hp generating station at the Falls.

Regardless of their party affiliations, Ontario newspapers with few exceptions favored public ownership. The city councils of Toronto and other Ontario municipalities were united in their determination to check further encroachments by the privately owned utilities. Only in the Provincial Legislature did the Syndicate wield dominant influence, so that in 1902, when for the third time the Toronto City Council asked for legislation authorizing the city to buy and distribute electricity, the Legislature's answer was again negative.

But in the Legislature, too, the public-ownership advocates, chiefly members of the opposition Conservative Party, were gathering strength. Unable to obtain favorable action on Toronto's petition, they sponsored the following resolution:

> In all future agreements made between the Commissioners of the Queen Victoria Niagara Falls Park and any other person or persons, power shall be reserved to the Provincial Government to at any time put a stop to the transmission of electricity and pneumatic power beyond the Canadian boundary; and that in the opinion of this House, the water of the Niagara River and its tributaries, as well as the waters of other streams, where necessary, should at the earliest moment and subject to existing agreement, be utilized directly by the Provincial Government in order that the latter may generate and develop electricity and pneumatic power for the purpose of light, heat and power, and furnish same to municipalities in this Province at cost.

When the roll was called, 30 Conservative members voted for the resolution and 41 Liberals against it.

In Toronto, however, both businessmen and political representatives continued to press the attack. The Toronto branch of the Canadian Manufacturers Association endorsed the City Council's demand for municipally owned power and set up a committee of well-known citizens to coöperate with the Council, under the chairmanship of P. W. Ellis, the manufacturing jeweller who had been instrumental in the creation of a similar committee by the Toronto Board of Trade. W. K. McNaught was also on the committee. Subsequently, both Ellis and McNaught were to serve on the Ontario Power Commission, an investigative body which preceded the establishment of Ontario Hydro.

Other members of the Manufacturers Association Committee were A. W. Allen, E. H. Gurney, A. E. Kemp and J. O. Thorn, all prominent businessmen. The Committee worked closely with the special Committee of the Toronto City Council and with representatives of the other municipalities that were now joining the fight, although the smaller industrial towns of Ontario had not wholly suspended their chronic distrust of Toronto, frequently referred to as "hogtown" because of its alleged tendency to shoulder aside smaller communities.

Alderman Spence and other Toronto leaders of the public-ownership movement helped to allay these suspicions by assuring the smaller communities that they would share fairly not only in the struggle for cheap power but in the benefits victory would bring. Now, in their advocacy of public ownership, some of the manufacturers in the Grand River Valley had become as zealous as Alderman Spence himself. Among the most vocal of these small-town leaders were E. W. B. Snider of St. Jacobs and D. B. Detweiler of Berlin, both descendants of Swiss immigrants who had come to Canada via the United States early in the nineteenth century.

With interests in flour milling, lumber, light manufacturing, and municipal traction, Snider was successful both as a businessman and as a politician. As the former, he was the first miller in North America to introduce the revolutionary Hungarian rolling-mill process, and as the latter he sat in the Ontario legislature from 1881 to 1894 as one of the members from Western Ontario. Detweiler, for his part, had ended his formal education after grammar school, entered business as a shoe salesman, and risen to a partnership in his company.

Sometimes singly and sometimes in combination, Snider and Detweiler have been hailed as the true fathers of Hydro: Snider as the architect who designed the imaginative system of municipal coöperation, Detweiler as the fervent prophet to whom the cultural and industrial gains to be won through public ownership became a personal obsession. For more than a decade prior to his emergence as its most impassioned advocate, he had ridden a bicycle over the dusty gravel roads of the Grand Valley, preaching to all who would listen of a day when Galt and other towns would have electricity to light their streets and homes, run their factories, and banish forever the drudgery of the woodpile and the inefficiency of the kerosene lamp. Scorned at the beginning, Detweiler fared better than most prophets in their own country. His bicycle is now a treasured exhibit of the Waterloo Historical Society.

Strangely enough, the alliance between these two men came about more or less by accident. In an address to the Waterloo Board of Trade, Snider urged that the town, like Toronto, should think about getting its power from Niagara, and to that end should seek the coöperation of the Boards of Trade of the nearby towns of Berlin, Galt and Guelph, and the mayors of Preston and Hespeler. Detweiler read a report of this speech in the Berlin *News Record* and wrote to Snider, receiving a reply in which Snider repeated substantially what he had said at the Board of Trade banquet in Waterloo.

The correspondence led to an evangelical alliance that functioned effectively during the three months that preceded the Ontario elections that spring. Addressing the Berlin Board of Trade, of which he was Vice-President, Detweiler urged that a committee be formed, composed of representatives of the inland towns which, like Toronto, had been frustrated in their hopes of getting Niagara power. He was strongly supported by the President of the Board, S. J. Williams, who suggested that the towns be asked to finance the committee with a $5,000 fund.

The Berlin businessmen agreed, and appointed Detweiler and Snider as a committee of two to prepare a resolution dealing with the matter. That was all the authority the two dedicated crusaders needed. The morning after the meeting, Snider wrote to Charles H. Mitchell, consulting engineer of the Ontario Power Company and later Dean of the School of Practical Science at Toronto University, asking if he would come to Berlin along with Alderman Spence to advise the Berlin Board of Trade how best to get action from the Provincial Legislature to advance the prospects of obtaining Niagara power for both Toronto and the smaller industrial towns of Ontario. When Mitchell replied a a week later he was careful to point out that, as an employee of a private utility, he could scarcely be considered a "disinterested engineer" —indeed at that time only a few electrical engineers in Ontario could be so considered. He stated, however, that he would come, speak, and charge a nominal sum—$25—and expenses. Alderman Spence also accepted.

Meanwhile the provincial elections had been held and showed a sharp swing away from the Liberal Government of Premier George W. Ross, which had so consistently said no to the advocates of public ownership. Its majority was reduced to one, but whether public ownership had been a decisive factor was open to question. Far more likely, it was generally believed, the voters of Ontario had finally grown tired of a government that had been continually in office for more than thirty years. Nevertheless, when Snider, Detweiler and Spence studied the election returns, they saw three newly elected members on whom they felt they could reasonably count: Adam Beck, a colorful personality who had twice been elected mayor of London, the largest city in southwestern Ontario; John S. Hendrie, mayor of Hamilton; and the belligerent eccentric, Dr. Beatty Nesbitt, whom cartoonists pictured declaiming as he raised an angry fist, "The Government at the switch, not corporations."

An all-day meeting with luncheon was held in Berlin on June 9, 1902, at which C. H. Mitchell and Alderman Spence were the principal speakers and for which twenty-five firms and individuals had subscribed among them $45 to meet expenses. The first name on the list was August R. Lang. Chairman of a public commission which had taken over the franchise of the Berlin Gas and Light Company, he had discovered on investigation that the cost of supplying Berlin with electricity would be twelve cents a kilowatt hour using manufactured gas, the cheapest fuel available: this charge was much greater than that proposed by the advocates of municipal distribution of Niagara power.

At the morning session, delegates from Toronto, Guelph, Preston, Waterloo, Hespeler and Berlin heard Mitchell review developments at Niagara and declare that power could now be transmitted one hundred miles to the manufacturing plants of mid-western Ontario at a cost of $17 per year per horsepower.

Alderman Spence's speech which followed was a carefully wrought masterpiece, embodying as it did a proposal to be presented to the Legislature which included all the essentials of the Hydro legislation that was to be subsequently adopted. What was required, said Mr. Spence, was "a government Commission which would have power to arrange for the transmission of electricity to the various municipalities desiring it; this commission to issue its own bonds in payment of transmission lines, which bonds would be covered by bonds of the municipalities interested. Under the scheme, the government, through a commission, would undertake the transmission to the municipalities desiring power, the latter guaranteeing by their bonds the cost, and selling in turn to all manufacturers at an even rate, preventing in this way the power from falling into the hands of any monopoly, and in this way securing to the industries of this Province advantages of cheap electrical energy."

In the afternoon, representatives of Guelph, Preston, Waterloo and other municipalities took the stage, endorsed the recommendations of the morning speakers and urged that an action committee be appointed to take further steps. Among the appointees were E. W. B. Snider and D. B. Detweiler, but their associates on this historic committee also deserve to have their names recorded, since in effect, they too were the founding fathers of Hydro. These were the names:

PRESTON: George Clare, S. J. Cherry, George Pattinson

HESPELER: George Forbes and W. A. Kribs, M.P.P.

WATERLOO: William Snider and R. Roschman

BERLIN: J. Lang, C. K. Hagedorn, D. B. Detweiler

ST. JACOBS: E. W. B. Snider

GUELPH: C. Kloepfer and L. Goldie

TORONTO: F. S. Spence and P. W. Ellis

GALT: M. MacGregor, Joseph Stauffer and R. Scott

BRIDGEPORT: P. Shirk

BRANTFORD: The Mayor, D. B. Wood and Lloyd Harris

Three weeks later the Committee held its first meeting. It was attended by only ten of the twenty-one members but in addition, Aldermen MacMechan, Cooper and Geary of the City of London were present as representatives of the newly elected Mayor, Adam Beck. This meant, in all probability, that Beck himself would soon come out in support of the movement.

The Committee appointed Alderman MacMechan to go with E. W. B. Snider and F. S. Spence to the Legislature and present the views of the coöperating municipalities. Detweiler was instructed to obtain from the municipalities estimates as to the amounts of power they would probably require, and from the power companies at the Falls the prices at which they would be prepared to supply these amounts. The required information was promptly supplied by fourteen municipalities: Preston, Hespeler, Baden, Berlin, Waterloo, New Hamburg, Galt, Ingersoll, Woodstock, Guelph, London, Stratford, Paris and Brantford.

No further meeting of the Committee was held until October 20 at Galt. Only nine members attended but the Galt manufacturers appeared in force. Most of them agreed with Snider that Premier Ross's Government was unlikely to undertake the provision of cheap power for the people; hence other steps should be taken. Would any of the power companies be willing to supply power at $15 per horsepower? The Committee agreed to investigate the possibilities immediately.

Not all of the Committee, however, were convinced that Premier Ross and his Government would maintain their opposition to public ownership. The Liberals had lost ground at the last election and the disaffection of the municipalities could have been a factor in the vote. Would not this induce the canny Premier to offer some kind of a compromise?

That would depend, it was concluded, upon how much weight in favor of public ownership the municipalities could bring to bear at the next session of the Legislature.

THE ISSUE JOINED

5

In January 1903 the cause of public ownership in Ontario, though now supported as a growing political issue, suffered what seemed to be a major setback. This was the announcement that the Queen Victoria Niagara Falls Park Commission had granted an irrevocable franchise to the Mackenzie-Pellatt-Nicholls Syndicate to generate 125,000-hp at Niagara Falls for a yearly rental of $15,000. As if to add insult to injury, the Commission further agreed to abstain from using water to generate power except for park purposes.

On this basis the Syndicate applied to, and promptly received from, the Liberal Government of Ontario a charter for the Electrical Development Company already mentioned. The fact that the Syndicate also controlled the Toronto and Niagara Power Company, chartered by the Dominion Government in 1902, placed the Mackenzie group in a position to exercise the rights granted by both the Parks Commission and the Federal Government, to exploit which the Electrical Development Company was capitalized at $11,000,000, of which $6,000,000 was in first mortgage bonds and $5,000,000 in common stock.

The success of these manoeuvers and the speed with which they were accomplished did nothing to enhance the Syndicate's reputation, much less its popularity, among the advocates of public ownership. Already embittered by repeated frustrations in the Provincial Legislature, the latter now raised their voices in angry protest. In the press and from the rostrum the Syndicate was pictured as a band of robber barons intent upon placing the people of Ontario in a state of perpetual economic bondage. Why, it was demanded, should a handful of avaricious men, notorious for their insolence in public dealings, reap unwarranted profits from the exploitation of a great natural resource which belonged rightly to the people? And how long, it was asked, would the politicians

36

at Queen's Park, the seat of the Provincial Legislature at Toronto, continue to deny the demands of the Toronto City Council and other municipal bodies for a share of the water power that had been given away so casually to a group of greedy, predatory and parasitical promoters? Thus in the public mind was created an image of a few privileged men, vigilantly and with great cunning and resource, holding the mass of ordinary citizens under tribute by arbitrarily supplying very poor service at very high cost.

The issue was now joined, and as public pressure mounted in the form of indignant protests and by way of resolutions passed at public meetings, Premier Ross, the leader of the Liberal Government, felt that the time had come for a compromising gesture. This took the form of a letter addressed to the Park Commission requesting an estimate of the cost of transporting power within a practicable distance of the Falls. The softening of a hitherto obdurate attitude to public ownership demands had been preceded by another gesture: a statement made by the Queen Victoria Niagara Falls Park Commission. Concurrently with the granting of the Electrical Development Company franchise, the Commission announced that "the plans of the three companies now exercising their franchises contemplate such a large output of electrical power that . . . there is no likelihood of anything like the demand being for many years equal to the supply, and consequently the tendency will be to compete for the business offered."

But the Toronto City Council, now obdurate in its turn, refused to pin its faith to any such hypothesis. Supported by the Toronto Board of Trade, it again asked the Legislature for authority to develop and distribute electric power—and for the fourth time was refused. The Council, it was claimed by its opponents in the Legislative Assembly, had "no matured plan" for the utilization of the power it requested. Moreover, it was contended, if Toronto built a transmission line of its own, the smaller municipalities would be at a disadvantage when the time came to provide for their own power needs.

These arguments ignored the coöperation which had been achieved by the municipal representatives who had convened at Berlin, and their firm alliance with Toronto. Equally specious was the contention of the private-power lobby that the plans formulated at Berlin would give an unfair advantage to manufacturers and industries located within a short radius of Niagara. The coöperating municipalities had anticipated this very objection and had answered it fully in their published plans, which were vigorously endorsed by the manufacturing and business interests, both in Toronto and in the smaller cities of the province.

Lest the Government be in any doubt on this point, the Toronto

Branch of the Canadian Manufacturers Association on February 13, 1903, forwarded to Premier Ross a most forceful resolution. The citizens of Toronto, it said, had no assurance that the private companies that had recently been given charters would charge fair prices. The Association therefore placed itself on record "as favorable to an immediate action on the part of the City before further vested interests are created, and pledges its support to any line of action by the City, either alone or in conjunction with other municipalities, that will guarantee electric power and cheap light to the citizens of Toronto for all time to come at the actual cost of the same, or at a fixed percentage of profit upon the cash expended."

Coming not from petulant populists but from an organization of businessmen noted for their caution and conservatism, this was a remarkable pronouncement. And while Premier Ross was still weighing its political import, the Snider-Detweiler Committee held yet another meeting in Berlin, this time in the YMCA hall. Attendance was large, representing nineteen municipal councils—Toronto, Berlin, Brantford, Bridgeport, Dundas, Galt, Guelph, Hamilton, Hespeler, Ingersoll, London, Preston, St. Catharines, St. Jacobs, St. Mary's, St. Thomas, Stratford, Waterloo and Woodstock. Boards of Trade and Manufacturers Associations were also present, as well as deputations and individual businessmen from ninety different Ontario communities. And for the first time, the redoubtable Adam Beck, whose attitude to public ownership had previously been obscure, appeared at the meeting in person. A successful manufacturer in his own right, Beck was both a member of the Conservative opposition in the Ontario Legislature and Mayor of London, the latter position won after a spectacular campaign he had waged for civic improvement.

The Berlin meeting was called to hear the Snider-Detweiler report that had been asked for at the October, 1902, meeting in Galt of the Committee of Manufacturers and Consumers of Motive Power interested in the Transmission of Electric Power from Niagara Falls. The report marks one of the pinnacles of achievement in the struggle for public ownership and is given in full here:

> Your sub-committee, appointed to investigate the practicability of this proposal to transmit electrical power from Niagara Falls to cities and towns within reasonable distance from the Falls, and the probable cost of power so transmitted, beg to report that the prospects are most encouraging for realizing the project within a comparatively short time, either through dealing with existing power companies for the required power,

or by direct efforts of the municipalities interested. Interviews have been had with officials of various existing companies and terms have been discussed with them and while we were not in a position to close any contract and therefore are not authorized to place before you any definite proposals from any of these companies, yet from conversations with their officers and from our knowledge of the conditions affecting the operations of the companies, we believe that we are justified in assuming that electric power can be obtained at a price, if taken in large quantities, of about $7 or $8 per continuous horsepower per annum delivered at Niagara Falls or from $14 to $15 per horsepower delivered to the various municipalities.

Of the two methods as suggested your committee recognized that the latter plan, the power company delivering the power to the various municipalities, is by far the less difficult to put into operation, while the former plan, under which the municipality would own and operate the means of transmission, has the advantage that in case at any time the municipality should desire to change their source of supply or to undertake by any coöperative scheme to develop their own power they would have in their own hands the means of transmission. It is perhaps unlikely that the occasion will arise in the near future that municipalities will desire to enter into the development of power, since existing companies will, within two years, be in a position to deliver far more power than is likely to be demanded for a much longer time; but the possibility is worthy of being kept in mind considering the method to be adopted of obtaining power.

To enable the municipalities to so purchase and sell or otherwise distribute electrical power, legislation would, of course, be necessary, and we recommend that prompt action be taken towards securing from Legislature at their approaching session the necessary powers enabling municipalities to undertake such work. Such legislation should empower municipalities to coöperate, when authorized by the vote of the property owners of their respective municipalities, to develop and transmit or distribute electrical energy; or to buy and transmit such power; or to buy power delivered at the several municipalities and to sell and distribute the same within their own limits. In case of municipalities coöperating in such work it would be necessary to devise means whereby the interests of the various municipalities would be protected, the rate for power kept uniform in the several municipalities and a scale of rates obtained which would prevent any municipality cutting rates against any other. In carrying out such a coöperative scheme, provision must be made for apportioning the cost among the various municipalities in proportion to population; and provision should be made for subsequent admission of other municipalities on the basis as to cost which would be fair to the original municipalities forming the union.

Should, however, a plan of municipal action in this direction fail to receive the approval of the Legislature your committee recommends that consumers take united action forthwith, and close arrangements with one of the existing power companies for transmission of electrical energy, either direct to the consumer or to local organizations in each town and city who would purchase the power in quantity and distribute to consumers in their several municipalities.

Your committee feel very strongly that to ensure success under either of the methods above proposed, and to obtain the best results, the united efforts of as large a number as possible of either municipalities or individual consumers is absolutely essential since companies prepared to dispose of power would be willing to give much better terms if large quantities of power can be disposed of in a single transaction than if only small quantities can be sold.

The report was the most comprehensive that had yet appeared and was enthusiastically approved, with an amendment calling for immediate action by the Ontario Government. Introduced by the mayors of two of the largest communities in the province, Urquhart of Toronto, a Liberal lawyer and long an advocate of public ownership, and Beck, the new Conservative recruit from London, the amendment cut across party lines and urged upon the Provincial Government the advisability of "building and operating as a Government work, a line for the transmission of electricity from Niagara Falls to the towns and cities; and that the municipalities here represented call upon their representatives in the Legislative Assembly of Ontario to urge upon the Government to carry out this resolution."

The genesis of the proposal outlining the anatomy of Ontario Hydro is unknown, but whatever may have been its inspiration, it set the course by the novelty, simplicity and practicability of its design, which the advocates of public ownership were to follow to their final victory.

Press coverage of the Berlin convention was extensive and much of it vigorously editorial. The Toronto weekly, *Saturday Night*, then one of the leading Canadian journals of opinion, had this to say: "If the members of the Ontario Government are wise, they will no longer deaden their ears against the rising clamor of the multitude. . . . 'Berlin Convention' . . . are words which will henceforth possess a new and a better significance. They will now connote a declaration of popular rights as opposed to monopolistic privilege. They will suggest a new, and let us hope, a better tendency in the administration of great public franchises. They will inspire with new hope and purpose all advocates of industrial and political progress, who believe that the many were not designed to be forever bled and bullied by the few." The words "bled" and "bullied" are indicative of public feeling against the private interests.

Ten days after the Berlin convention, a delegation from the participating municipalities, with Snider and Toronto Alderman Spence as its spokesmen, met at Queen's Park. They asked Premier Ross whether the Government would undertake the work of providing power at cost. If not, then the municipalities should be allowed to do it for themselves. Had not the Ross Government already spent large sums in developing the mineral resources of New (Northern) Ontario? There could be no question, therefore, of sectional favoritism or advantage. Then why deny to the people the great and inexhaustible resource of Niagara? What was needed, they asserted, was a Niagara Power Commission with authority to buy and transmit power and, if necessary, to develop it at the source as well.

Replying, Premier Ross repeated the argument already advanced by the Niagara Falls Park Commission and complacently endorsed by the private power companies: that Niagara power would be in surplus supply for twenty years or more to come. Implicitly, in the very nature of things, fair rates would be automatically provided by the workings of competition. Nevertheless, the Premier promised on behalf of his Government that a bill would be introduced enabling Toronto and other municipalities to carry out the coöperative plan proposed by Snider. Shortly thereafter, in a speech at Newmarket, a small manufacturing and marketing town some thirty-five miles north of Toronto, Ross outlined his cautiously balanced policy. The Province, he pointed out, would continue to get a revenue of $300,000 a year from the three private companies with development rights at Niagara Falls; then he continued: "We are willing to allow municipalities, Toronto and the rest, to develop energy there and they will not be curtailed. But Ontario must not go into debt because of it. Niagara power can only reach a small proportion of the population of the Province. All and any municipalities desiring to go into the business of developing electric power may do so, but the Government will not involve this Province in debt unless for the substantial benefit of all."

While this was more than the Premier had previously conceded, it did not satisfy the leader of the Conservative opposition, James P. Whitney, who also spoke from the same platform. It was impossible to justify, said Whitney, either the granting of a third franchise to private interests or the unrestricted export to the United States of power developed on the Canadian side of the Niagara River. The Government, he insisted, "should investigate the problem of supplying power to all places within 150 miles of Niagara, or appoint a Commission to provide the means of enabling the Union of Manufacturers to do so."

As both the Opposition leader and the Premier were aware, the municipalities were ready to campaign for what they wanted, and had agreed upon an equitable distribution of the costs of such a campaign: $25 each from places with less than 5,000 population; $50 from places of from 5,000 to 10,000; $75 from those with 10,000 to 20,000. Hamilton and London had each agreed to pay $125, and Toronto $400.

Confronted by this threat, the Premier acceded to the demands of the Opposition and in May appointed a fact-finding commission, with Avern Pardoe as Secretary. Its report provided the first comprehensive compilation of data on the progress of municipal ownership of public utilities. Some 126 municipalities in Ontario, it was reported, were then operating their own waterworks, electric light and gas works, the majority of these services having been taken over within the preceding ten years. Each municipality had acted independently, so that the obvious administrative advantages of coöperation were wholly lacking.

On May 7 Premier Ross introduced his Power Bill, which was passed in amended form on June 12. It fell considerably short of what the Berlin Committee had finally demanded, namely government construction and operation of a transmission line from Niagara to the coöperating municipalities. In other respects, however, the Bill, entitled "An Act to Provide for the Construction of Municipal Power Works and the Transmission, Distribution and Supply of Electrical and Other Power and Energy," gave the public-power advocates pretty much what they had wanted. Any two or more municipalities now had the right to acquire or construct works for the generation and distribution of electric or other power and energy. But they were required, before starting work, to undertake a full investigation of the desirability of the project. To carry out this investigation the coöperating municipalities were to appoint a Commission, which was to consist of one electrical engineer and at least two, but no more than four, other persons, professional or business men, to report on the desirability of the project, its probable costs, and the proportion of cost to be borne by each of the contracting municipalities.

Upon acceptance of the report the municipalities could then proceed with the work, under another Board of Commissioners to be appointed by the Chief Justice of Ontario. This Board was given very wide powers to buy or build, and to fix the rate or price to be charged for delivered energy. But the Commissioners were not to take action until their proposal had been submitted to the qualified electors (property owners) and authorized under a by-law. If this approval by the electors were forthcoming, then the Board was empowered to find the needed money by issuing debenture bonds within certain limits. They could not increase debenture debt beyond the amount fixed by the Municipal Act, nor was any municipality permitted to increase the rate of taxation which it could levy under the act.

On July 9 the municipal coöperators met in Berlin to appraise what they had won and to implement the authority granted to them. His mission accomplished, D. B. Detweiler, pioneer zealot of the movement, handed in his minute book and his resignation. He was succeeded as Secretary by J. C. Haight, a Waterloo barrister. Yet another of the Berlin series of committees was appointed to gather information and take further steps toward drawing closer together the Union of Municipalities comprised of those communities which wanted publicly owned power. By August 12 the indicated steps had been taken and representatives of sixteen municipalities met in the Toronto City Hall to set up an investigating commission under the provisions of the Ross Power Act.

The Commission, which became known as the Ontario Power Commission, consisted of E. W. B. Snider, Chairman; P. W. Ellis, Vice-Chairman and Treasurer; J. C. Haight, Honorary Secretary; Adam Beck of London and W. F. Cockshutt of Brantford. The firm of Ross and Holgate of Montreal was engaged as consulting engineers and Professor R. A. Fessenden of Washington, D.C., was appointed as the engineer-commissioner required by the Act. John McKay of Toronto was appointed financial adviser. All the Commissioners acted without remuneration except Professor Fessenden, who was paid a nominal fee, the consulting engineers charging only their costs. To finance the Commission, the Toronto City Council voted $11,756 and that of London $1,542. The remainder was pro-rated among the other coöperating municipalities. Of the original sixteen municipalities, nine dropped out under the pressure of hostile influences. The remaining seven—Toronto, London, Brantford, Guelph, Ingersoll, Stratford and

Woodstock—held an organizing meeting in Toronto on December 8, 1903.

Under the chairmanship of E. W. B. Snider, the Ontario Power Commission, the first of three commissions concerned with hydro-electric development, conducted a comprehensive study which took nearly three years to complete. For the purpose of gathering data about power consumption the services of Arthur V. White were enlisted: later he was to become consulting engineer to the Canadian Commission of Conservation. The task of White and his staff was to go from factory to factory and determine whether or not manufacturers would shift from steam-generated systems to hydro-electric power when and if Niagara power were offered to them.

Not all of the manufacturers were disposed to scrap their investments in steam plants, which provided a valuable by-product in the form of steam heating for the factories during the winter. The Premier's investigators also reported that a good many factory owners had tied themselves up in long-term contracts with the Electrical Development Company which had purchased rights-of-way for a transmission line from Niagara to Toronto around the head of Lake Ontario. While this information was being collected and placed before the Premier, he himself became involved in new negotiations with the Electrical Development Company. The Mackenzie engineers had discovered, in excavating for the forebay of their plant, that it would provide much more water than was stipulated in the rights granted the Canadian company by the Queen Victoria Niagara Falls Park Commission. Application was then made to the Park Commission for permission to generate 125,000-hp by the installation of ten 12,500-hp generators. The right to generate this power was embodied in an option contained in their original franchise. But the exercise of this option, it will be recalled, had been conditioned by the obligation to reserve one-half of the generated power "for the use of municipalities, for the purpose of operating a municipal system of lights, heating, or other public utilities."

An agreement embodying this proviso was prepared for submission to the Ontario government but was stalled by the company, which was unwilling to have any restrictions placed on the markets it might seek for all its generated power. Line transmission technology was improving and the radius of its potential market was expanding correspondingly. The company wanted all of the new power and obstinately refused to settle for less, despite political storm warnings that might well have dictated a less intransigent policy.

During the provincial election campaign conducted late in 1904, General Manager Nicholls argued vigorously to win special consideration for the Electrical Development Company as the one power firm at the Falls that could be called Canadian. The electors listened skeptically. The source of the capital investment, they felt, was less important to them than the power rates that might be expected if the company were to be given a virtual monopoly.

Premier Ross entered the campaign with confidence. The Liberal Party had been in office in Ontario for thirty-four years, the longest period ever attained in Canada and there seemed to him no reason why that condition should not be continued. The Ross Power Act assured his re-election. It was, in fact, an honest and statesmanlike resolution of the issue. Unhappily for Mr. Ross, however, it had been achieved too late and under much too embittered circumstances.

HYDRO IS BORN

6

In the first round of the fight in Ontario for power at cost, the advocates of public ownership wrung from a reluctant Liberal Government the formal authority for municipalities to engage in the purchase or development and distribution of electric power, providing the municipalities themselves could finance such undertakings under the limited powers granted them by the existing Ontario Municipal Act. In the second round the Liberals were swept out of office after a tenure of thirty-four years, to give the Conservative Party a majority of 40 in a 98-seat unicameral assembly. While many other issues were involved in the campaign, the overwhelming defeat of Premier Ross enabled the public-ownership proponents to claim the election results were tantamount to a declaration that what had been given them was not enough. Had not the new Premier, James Pliny Whitney, promised them much more in his campaign speeches? The water power of Niagara was the property of the Canadian people, he had declared, and the Provincial Government must help them to enjoy their heritage without paying tribute to profit-seeking interests.

No sooner had the new Conservative Government taken office than Whitney proceeded to make an immediate down payment on his promises. Repeating a campaign statement that "the water power of Niagara should be as free as air," he added, "and more than that, I say on behalf of the Government, that the water power all over the country should not in the future be made the sport and prey of capitalists and shall not be treated as anything else but a valuable asset of the people of Ontario, whose trustees this Government of the people are." The stalled agreement with the Electrical Development Company for an additional allocation of horsepower was cancelled and the Premier announced that no further franchises would be granted at Niagara until the whole question had been thoroughly examined.

Adam Beck had again been elected member for London by a substantial majority and was appointed a Minister without Portfolio, though in effect Minister of Power, in the Whitney Cabinet. He became the Government's principal spokesman on power matters, and soon after his appointment made a speech in the Legislature even more challenging than those he had given during the campaign. Deriding the provisions of the Ross Electric Power Company agreement as not worth the paper they were written on, he was equally scornful of the preposterous assertion that because three competing companies had been given franchises at the Falls, the public would be safeguarded against rate-fixing. In Montreal, Beck pointed out, there were two competing companies until one absorbed the other—after which lighting rates jumped from 12½ to 14½ cents per kilowatt hour, and motor service rates even more. All the private companies, said Beck, were cut to the same pattern: "The promoters get the capital stock for nothing, the total cost of buying and developing the property being borne by the proceeds of the bond issue."

Concerning the Ross Power Act, Premier Whitney and Beck agreed that by making the financing of coöperative enterprises the unassisted task of the municipalities, the Act in effect evaded the proper responsibilities of the Province. Without the backing of the Provincial treasury the municipalities would almost certainly fail. Moreover, the Act made no provision for an administrative authority empowered to integrate the several municipalities in an effective coöperative whole. Under the circumstances, the Premier and his minister decided it would be better to repeal the Ross Power Act and pass a better one.

Accordingly, on July 5, 1905, five months after taking office (February 8, 1905), the Government issued an Order-in-Council creating the Hydro-Electric Power Commission of the Province of Ontario. Again, it was to be a commission of inquiry, and as such could be set up by the Cabinet without reference to the Legislature, thus avoiding debates and delays. Adam Beck was appointed Chairman. His two fellow-Commissioners were P. W. Ellis and George Pattinson of Preston, a member of the Provincial Legislature for South Waterloo. They chose as their chief engineer an able young Canadian, Cecil B. Smith, who from 1899 to 1901 had been Assistant City Engineer in Toronto, and resident engineer of the Canadian Niagara Power Company from 1901 to 1904. Smith had participated in the discussions between the City of Toronto and the power companies and was credited with having helped to mobilize the municipalities against the encroachments of the Electrical Development Company.

Under Adam Beck's chairmanship, the newly constituted Hydro commission of inquiry began a comprehensive survey of the developed and undeveloped water resources of the entire province. For this purpose Ontario was divided into five districts:

1. *Niagara*, including the populous region from Toronto westward to the Detroit and St. Clair Rivers.
2. *Trent*, the area embracing the watersheds of the Trent and Moira Rivers flowing into the Bay of Quinte near the eastern end of Lake Ontario.
3. *Georgian Bay*, covering the area immediately north of the Niagara district drained by the rivers flowing west to Georgian Bay and Lake Huron.
4. *Eastern*, comprising the watershed of the Ottawa River in Ontario, together with the eastern counties between the Ottawa and St. Lawrence Rivers.
5. *Northern*, the Algoma, Thunder Bay and Rainy River districts, a vast and sparsely populated territory containing the greater part of Ontario's forest and mineral wealth.

While the earlier commission, appointed by the Ross Government and headed by Snider, confined itself to Niagara and its environs, both commissions sought essentially the same data: the hydro-electric potential of the available water power; the capital cost of installing plants and transmission lines; the present and future demand for electric power and the rates which would have to be charged.

Most of the private electrical companies refused information about their costs and rates. This was within their rights, except in the case of companies under contract with the Queen Victoria Niagara Falls Park Commission, which were obliged to make semi-annual statements of the power generated and sold. Unfortunately the Commission had never enforced this requirement. By this time thoroughly frightened, other private companies felt that they should not be asked to supply the noose with which they were likely to be hung. But they failed to maintain a united front. The Canadian Niagara Power Company quoted prices to the Snider and Beck Commissions, with the result that subsequently the Electrical Development Company found its bargaining position badly impaired.

After serving on both the Snider and the Beck Commissions during 1905, Ellis resigned on January 26, 1906, from the Beck Commission and was replaced by John Milne of Hamilton. It is probable that Ellis resented Beck's cavalier treatment of Snider, who had fought for public power long before Beck had made up his mind to enter the lists. By a coincidence that Snider never forgave, much of the information contained in the report of his Commission, which was released on March

28, 1906, appeared in the newspapers of the same day, credited to Adam Beck. The investigations of the Beck Commission covered much of the same ground, but this was not the first nor would it be the last time that Beck was to manifest his singular capacity for alienating friends and making enemies.

In the report of the Ontario Power Commission, Snider and his colleagues recommended that the seven municipalities for which they spoke—Toronto, London, Brantford, Guelph, Ingersoll, Stratford and Woodstock—proceed with a 60,000-horsepower development unless other municipalities along or near the line joined them. There were eleven of these: Hamilton, Berlin, Dundas, Galt, Hespeler, Mitchell, Paris, Preston, St. Mary's, St. Thomas, Waterloo. If these eleven should also become committed to the project, then the Commission's recommendation would be for a 100,000-horsepower development. The relative capital cost of setting up the two systems was shown to effect a saving of several million dollars for the larger installation, since the cost of the 60,000-hp system with transmission lines was estimated at $9,000,000 as compared with about $12,000,000 for the 100,000-hp system. The estimated saving in annual carrying costs was even more striking.

Existing distributing companies would have to be purchased, including steam plants in Toronto, London and Brantford, the Commission reported, with the recommendation that the municipalities apply for authority to do this. To finance the program, each of the participating municipalities would then issue debentures to the amount of its share of the capital cost. These debentures would be sold by trustees, the proceeds to be paid to a board of Construction Commissioners appointed by the participating municipalities. The Construction Commissioners would issue bonds to the amount of the debentures.

As if to forestall the expected attack by the private-power interests, the Commission concluded its report with this forthright declaration:

> The basal fact that power and light can be supplied under a municipal development, properly carried out—under engineering conditions equal to those of its commercial competitors, at prices beyond the reach of permanent commercial competition, is not open to argument. . . . Competing (private) companies have to pay considerably higher interest rates on their bonded debt, and in addition they have large issues of capital stock on which dividends have to be earned. Whether rates be fixed by the companies voluntarily or under Government regulation, regard must be paid to these conditions and the rates loaded

accordingly. No criticism directed at detached isolated facts or figures will alter these broad underlying conditions, from which the general public will derive benefits otherwise unobtainable.

On April 4, a week after Snider and his colleagues released their report, Beck began publication, in three instalments, of the report of his commission of inquiry.

The two commissions were in substantial agreement with respect to the demands for power and the savings to be achieved by a prudently administered publicly owned system, but there were important differences in certain of their recommendations. The Snider Commission recommended that the municipalities construct at Niagara a generating plant of 60,000-hp or more on a fourth site in Niagara Falls Park. In addition to this installation, which it estimated would take two or three years to complete, the Commission recommended the construction of transmission lines and local distributing systems. The Beck Commission, on the other hand, estimated that four years would be needed to build a generating plant at the only acceptable site available. It recommended, therefore, that transmission lines be built, but that power be purchased from the private company "which had already a supply exceeding present Canadian demands."

A week after the release of the first instalment of the Beck report, a public meeting in Toronto was told that power users in Fort William were paying $15 per horsepower per year and that the municipally owned hydro-electric plant in Orillia was supplying local manufacturers with current at $16 per horsepower. These figures tended to support the estimate of the Snider Commission that even with the minimum 60,000-hp installation, a coöperatively owned and operated system could supply motor service at from $15.73 per horsepower per year in Toronto to $23.87 in London, the city the greatest distance from the proposed generating plant at Niagara. All these rates were much lower than the rates being charged by private companies. For example, the Electrical Development Company was charging the Toronto Electric Light Company and the Toronto Street Railway $35 per metered hp per year, or a flat rate of $25 per year for 24-hour hp.

The day after the Toronto meeting, 1,500 delegates from 70 municipalities, from Kingston in the east to Sarnia in the west, assembled in the Toronto City Hall. Under the stimulus and tutelage of Adam Beck, whose hand in the organization of this massive demonstration was scarcely concealed, the delegates passed a resolution demanding Hydro legislation which closely resembled the provisions of a Bill which Beck and his advisers were already engaged in drafting. The text of the resolution follows:

That the municipalities now present and represented in the City Hall, Toronto, having an urban and rural population of over one million, respectfully urge upon the Lieutenant Governor-in-Council of the Province of Ontario the necessity of safeguarding the peoples' interest by originating as a Government measure legislation enabling the Lieutenant Governor-in-Council to appoint a permanent provincial commission with power to take, where considered by it advisable, the following action: The construction, purchase or expropriation of works for the generation, transmission and distribution of electrical power and light; to arrange with any existing development company or companies for power at a reasonable price, so as to be transmitted and sold by the Government to municipalities or others; also to vest in it the powers necessary to enable it to regulate the price at which electricity can be sold to all and every customer, whether municipal, corporate, or private.

Having enthusiastically endorsed the resolution, the delegates, 1,500 strong, formed in ranks of four and, with banners flying, marched to the Parliament Buildings in Queen's Park and thronged into the legislative chamber. With minor variations each delegation chairman chanted the same refrain: We want cheap power and we want it now.

In response to the insistent chorus the Premier reminded the delegates that his Government was pledged either to generate the power they demanded or to control the output of the private companies. But before he could sanction the expenditure of the $15,000,000 which the proposed program involved, careful consideration would be required.

In point of fact the experts of the Beck Commission had not been wholly in agreement. One of them, J. Stanley Richmond, published a letter in the April 18 issue of the Toronto *Globe* asserting that the cost of current at Toronto would be $27.73 per hp per year, not $16.53 as estimated by the Commission. And at a carefully staged press conference Adam Beck had denounced this heresy and sketched for the newsmen an eloquent vision of the future of a great, publicly owned Hydro for which he was prepared to fight the wicked Vested Interests to the death, if necessary.

The angry clash between Beck and his dissident colleague reflected both the brittle temper of the political contest and the lines on which it had been drawn: on the one side, the converts to public ownership passionately engaged in what had become an ideological crusade against the evil powers of monopolistic capitalism; on the other side, the beneficiaries and protectors of legally acquired property interests using all the weapons they could command to persuade the uncommitted opinion of the Province that public ownership could lead only to political patronage, graft, inefficiency, provincial bankruptcy, and ultimate disaster—in short, to an era of anarchy such as followed the collapse of the Roman Empire.

While tempers flared and charges and counter-charges won banner headlines in the Ontario press, Adam Beck was busily at work drafting the Hydro Bill. In this he had the expert assistance of Sir William Meredith, who had been leader of the Conservative Party while it was still in opposition, and was now Chief Justice of the Common Pleas Division of the Ontario High Court of Justice. Many years before, when the idea that power could be produced profitably at Niagara Falls was generally regarded as a kind of lunacy, Meredith had helped the Canadian Niagara Power Company to push through the Legislature the monopolistic franchise granted to it in 1892 by the Queen Victoria Niagara Falls Park Commission. Now, as the trusted adviser of the Conservatives, his great legal abilities were devoted to the drafting of a somewhat different instrument of state.

On May 7 Beck brought in his Bill, entitled, "An Act to Provide for the Transmission of Electric Power to the Municipalities." In introducing the measure the Minister without Portfolio spoke with unexpected moderation. Hydro, he patiently explained, intended to do business with the private companies, but on terms that would assure to customers in Canada rates as low as those being charged in the United States. A group of capitalists had assured the Government that they were prepared to deliver power at Toronto for $17 per hp per year, the figure mentioned in the Beck Commission's report.

The restraint of Beck's presentation was not adopted by his leader. If the private companies refused to accept this rate, said Premier Whitney, then the Government, as authorized by the Bill, would exercise its power of expropriation. He hoped and believed, however, that this would not be necessary.

The Bill created an administrative body called the Hydro-Electric Power Commission of Ontario, to consist of three members, two of whom might be and one of whom must be a member of the Provincial Cabinet and two of whom would form a quorum. The Lieutenant-Governor was empowered to appoint the Chairman and members to hold office at his pleasure, and salaries were to be fixed by the Government. The Commission was to appoint its own engineers, officers, and staff and fix their salaries.

Acting with the same powers enjoyed by the Minister of Public Works, the Commission could acquire by purchase, lease, or otherwise, any needed lands, waters, water power and power installations and make contracts with any corporation or private person for supply of power. They could also enter upon private property, and remove trees and other obstructions, in order to carry on the work of transmission and distribution. Any municipality requesting power would be told by the Commission the maximum price per horsepower at which it could be supplied. Plans and specifications, estimated cost of transmission lines, and distribution works would also be provided.

The municipality might then make a provisional contract with the Commission to supply its power needs, but this contract would not become permanent and binding until it received the approval of the ratepayers. The municipality was not to be ruled in these matters by the limitations of the Ontario Municipal Act, but the cost of any works undertaken by the Commission in order to supply a municipality with power was to be repaid to the Commission. The Commission might supply power to corporations such as railway companies, using the railway's right-of-way for the purpose. Any profit was to be used for the general maintenance expenses of the Commission.

The Commission was to fix the rate to be charged manufacturers, and also the rates which generating and distributing companies might charge. The price per horsepower to municipalities was to be at cost to the Commission, but the municipalities were to pay their share of certain charges: 4 per cent interest on capital account; an annual sum toward a sinking fund which would, in thirty years, retire the Provincial securities issued to finance the project; and cost of line loss and of operating, maintaining and insurance of the works.

The Province would guarantee loans raised for the purpose of the Act and the Commission would account to the Government for all money received. Any complaints by or against municipalities, or about unfair rates, were to be made to the Commission; and the Commission's decision about them was to be accepted under penalty of $100 a day. Any company refusing to give required information to the Commission was subject to a similar penalty. Neither the Commission nor the Province could be held liable for any error in estimates or specifications of works furnished by the Commission, and no suit could be brought against the Commission or any member of it in his official capacity without the consent of the Attorney General.

The Hydro-Electric Power Commission of Ontario differed radically from the commissions previously authorized. Under the Ross Power Act of 1903, various municipalities could unite in separate groups to work through separate commissions. Under the new Act all municipalities were required to make their wants known to the Commission and be guided by it into the most suitable of the local systems, all of them integrated into a general Provincial system.

The Bill was given its second reading on May 7 and on May 14 received Royal assent. Three weeks later, on June 7, the Whitney Government announced the membership of the Commission. Two were ministers of the Government: Adam Beck and the Hon. John S. Hendrie, Mayor of Hamilton. The third Commissioner was the engineer, Cecil B. Smith.

The Premier could scarcely have chosen two appointees more antagonistic in temperament than Beck and Hendrie. The former, whose grandparents emigrated from West Baden to Pennsylvania and thence to the "Pennsylvania Dutch" settlement in the Grand River Valley in Ontario, was forceful and opinionated, a martinet in business management and a prima donna in politics. A poor student, he had proved himself to be an apt apprentice when put to work in his father's foundry. During his twenties he built a successful cigar-box business. His brief political apprenticeship included an initial setback when he ran unsuccessfully for the Provincial Legislature. Subsequently he was twice elected Mayor of London as well as to a seat in the Provincial Legislature.

Colonel John S. Hendrie was a shrewd and cautious Scots-Canadian, as reserved in manner as Beck was flamboyant. Whitney had appointed Hendrie to the Commission, said the club gossips, to harness Beck while Beck harnessed Niagara. Hendrie, in fact, had never in the past displayed any enthusiasm for public ownership and had many private-power company supporters among his best friends.

Cecil B. Smith, the engineer member of the triumvirate, was ill-equipped to play the role of moderator. From a teaching post at McGill University he had gone to Toronto as the City's assistant engineer, then to the Canadian Niagara Power Company, and subsequently had managed the government-owned Temiskaming and Northern Ontario Railway. As chief engineer of the Commission of Inquiry under Beck he had been chiefly responsible for assembling the complex data on which the Commission's report was based. It was Smith who recommended the appointment of two young graduates of the University of Toronto, Fred A. Gaby and Harry Acres, who were chiefly responsible for the technical excellence of Hydro's operations in subsequent years.

With the elevation of Smith to membership on the Commission it became necessary to find a new chief engineer. For this critical post Beck chose P. W. Sothman, a Danish-born engineer who had constructed and managed a number of electrical installations in Europe and Africa. Twelve years before joining the Ontario Hydro staff Sothman had built a 25,000-volt transmission line in Silesia, the highest voltage used to that time. Later he had constructed generating plants in Johannesburg and had managed the Strasbourg electrical system which supplied current to a network of neighboring municipalities.

As an administrator Beck drove his associates as hard as he had driven the employees of his box factory. When they failed to match his own phenomenal capacity for work he regarded them as slackers and so informed them. Intolerant of political favoritism and nepotism, he demanded not only loyalty to the job but loyalty to himself as Hydro's boss.

The sole criterion of a person's acceptability was fitness for the job and efficiency in its performance. Beck also extended this principle to the choice of materials and equipment—fitness for the job. These qualities did not always endear Adam Beck to his associates but without them it is doubtful that he would ever have built the first and greatest publicly owned power system in the world.

THE DECISIVE YEARS:
1906-1908

7

The introduction on May 7, 1906, of the Hydro Bill in the Ontario Legislature—properly "An Act to Provide for the Transmission of Electrical Power to Municipalities" (6, Edward VII, chap. 13)—aroused the proponents of private ownership as nothing else had done. Heretofore, the conflict between the opposed factions had been largely ideological. Certain municipalities had established action committees, it is true, but these, like the commission of inquiry created by Order-in-Council in July 1905 had been largely investigative and discussion bodies. With the introduction of the May 1906 Hydro Bill as a government measure, notice was served on all and sundry that the period of investigation and polemics was terminated, and that the scores of resolutions adopted by representative bodies throughout the province in favor of public ownership were finally to be implemented, at least in part.

The response of the private-utility interests was as immediate as it was ill-advised. On May 8, the day following the introduction of the Hydro Bill in the Legislature, the Electrical Development Company celebrated the laying of the cornerstone of its new powerhouse at Niagara Falls. In performing the ceremony, Sir William Mortimer Clark, then Lieutenant-Governor of the province, noted that his official position precluded him from engaging in political controversy; nevertheless, he expressed his sincere trust that the gentlemen who had promoted the Electrical Development Company would be "permitted to enjoy the legitimate fruits of their enterprise."

Other speakers, free from the niceties of protocol, were more forthright—notably E. B. Walker, President of the Canadian Bank of Commerce, and W. R. Brock, a well-known Toronto manufacturer. The banker spoke of the country's good fortune in possessing men such as those of the Syndicate, with "expert ability and enterprise for the carrying out of important work." He then added, "these are not the kind

56

of works which it is given to Commissions and politicians to carry on
. . . these men make capital, see opportunity here for investment. And
yet the people who admire them would like, without their expert ability
and enterprise, to realize the same thing. We know that this kind of
public ownership, applied to a problem so intricate, is as useless as the
breath of the politicians who advocate it." Brock in turn derided "little
politicians who (have gone) around the country trying to make political capital out of the cry against them, but the best sense of the people
of Ontario (will) still stand by the men who took hold of schemes full
of great risk but with great possibilities for the country, and carried
them to a successful issue."

The indiscretion of the Lieutenant-Governor, coupled with the in-
flammatory remarks of the other speakers, provoked a storm of protest
throughout Ontario that caused Premier Whitney "to shoulder all
responsibility for His Honor's utterances." Nevertheless, on May 11,
the Toronto *Mail and Empire*, the principal Conservative Party organ
in Ontario, joined in the fray with comments on the cornerstone-laying
ceremony. "Those who described Mr. Beck's Power Bill as hostile to
capital can have made but a very hasty examination of it," read the
article. "A closer acquaintance with it will convince even those pre-
disposed to adverse criticism of the sincerity of the intentions of all
concerned. And we believe those intentions will be realized by the
provisions of the Bill. It (will) turn out as it was designed to be, a
measure of symmetrical reform—equitable towards power producers,
transmitters, distributors, consumers and the Province."

Continuing, the article recalled the recommendations made by the
early Snider Commission for public ownership of production, transmission and distribution, and contrasted them with the provisions of
the bill then before the Legislative Assembly. Beck, it stated, "did
not . . . adopt the recommendations to proceed without regard to
existing power companies . . . the Beck Bill does not thus stride over
the claims of existing companies. It leaves the private companies the
opportunities to do all the power business on the market; and so long
as they are satisfied with a reasonable profit and deal justly with all
customers, the Hydro-Electric Power Commission will not interfere
with them. It will leave them in possession of their franchise, their
works and their power. So long as they adhere to fair rates it will have
no cause to carry its regulative functions to the length of expropriat-
ing any power company's plant. But it is to be the judge as to what
rates are fair."

With the exception of the financial press, the same reaction was
reflected in a majority of Ontario newspapers. Nevertheless, the private-

power interests, led by the Mackenzie Syndicate, stepped up their campaign of ridicule and vilification against Beck and "the dreamers," "visionaries," and "crackpots," of which he was now the incandescent leader. The attack took on new force when the stock market, following the passage of the Hydro Bill, registered a sharp decline in private-utility stocks. Their bonds were heavily discounted also by the prospect that the municipal plebiscites to be held New Year's Day, 1907, would approve the local by-laws required to implement the Hydro legislation. The reaction of the investing public could have been expected since one of the principal charges laid against Beck was that he had based his cost estimates on "cooked figures." The utilities, it was claimed, would be ruined if they were forced to sell power at the ridiculous prices promised the municipalities by the Government's power minister.

Beck waited until July 24 to deal with these charges. The opportunity was created by a rally held in Galt when representatives of nineteen municipalities met with the directors of the Western Ontario Municipal Niagara Power Union to take appropriate action under the terms of the new Hydro Bill. The meeting was attended by both Beck and Commissioner Cecil B. Smith. Hendrie, the other Commissioner, was conspicuous by his absence, for which Beck apologized as he was to find himself doing many times in the future. Addressing the meeting, Beck adopted a tactic which later became characteristic in refuting charges laid against the Commission. The estimates in the report on which the Hydro Act was passed, he pointed out, were not his but those of competent and reputable engineers who had prepared them. Smith then made a detailed rebuttal of the charges, answering allegation and innuendo with facts and figures. Hydro's proposed rates, he said, were not 7½ cents per kilowatt hour, as had been alleged by the manager of the Toronto Electric Light Company, but 5 cents for residential lighting and 6 cents for commercial lighting, in contrast with the company's rates of 8 and 12 cents respectively.

A spokesman for the private-power interests was present in the person of the Hon. James Young, but after listening to him and other speakers, the delegates passed a resolution reaffirming their determination to obtain the benefits offered by the Hydro-Electric Act. An executive committee of formidable weight was appointed, consisting of the mayor or President of the Board of Trade of a number of the coöperating municipalities, to carry on the work of the Union. These included Mayors Coatsworth of Toronto, Clare of Preston and Butler of Woodstock; and for the Boards of Trade, Aldermen Fryer of Galt and Lyon of Guelph, Detweiler of Berlin, Woods of Brantford and

Matthews of London. The creation of yet another public-ownership action committee was in anticipation of the battle for public opinion still to come.

Meanwhile, private utility companies on both sides of the Niagara River were proceeding with their developments and planning future expansions which would require the withdrawal of so much water as to threaten seriously the beauty of the Falls. This threat assumed critical proportions when the Niagara, Lockport and Ontario Power Company applied to the New York State Legislature for permission to divert water to generate 400,000-hp.

Alarmed by this prospect, the Queen Victoria Niagara Falls Park Commission pointed out that the existing franchises, when fully exploited, would result in the diversion of 23 per cent of the entire flow over the combined American and Canadian Horseshoe Falls. Newly strengthened by the appointment of P. W. Ellis of early Hydro fame and Colonel L. Clarke Raymond of Welland, the Canadian park commission strongly urged that an international commission be created with authority to deal with the complicated problems of controlling Niagara's water for both scenic and hydro-electric purposes.

Even at this relatively late date (1906), the diversion of water for power purposes was governed by the statutes of the Ontario and New York Legislatures, and by the regulations of the complementary commissions set up in the 1880's to preserve the natural grandeur of the Falls and do away with the parasitical abominations which then afflicted the tourist trade. It had long been evident, however, that neither the two legislatures nor the park commissions they had created had the power or legal and technical knowledge to deal with the multitudinous problems involved in the control of boundary waters. Obviously, these were matters which lay within the jurisdiction of the Federal Governments of Canada and the United States and could be dealt with by them alone.

Apart from the eighteenth- and nineteenth-century treaties dealing with the international boundary and navigation rights on inland waters, the first suggestion that an international body be established to deal with matters not covered by existing treaties was made in 1895. That year, at a meeting of the Irrigation Congress of the United States held in Denver, Colorado, a Canadian delegate proposed a resolution asking that steps be taken for the appointment of an international commission, to act in conjunction with the authorities of Mexico, the United States and Canada in adjudicating the conflicting rights which had arisen, or might thereafter arise, on streams of an international character. The suggestion was approved in Ottawa and in 1896 the British

Ambassador at Washington informed the Government of the United States that Canada was ready to coöperate in the formation of an international joint commission.

There then ensued a hiatus until the passage of the Rivers and Harbors Act by the U.S. Congress in 1902, which asked that the President invite Great Britain to join in the formation of a commission to be made up of three representatives from the United States and three from Canada. As a result the International Waterways Commission was created in 1903. Its functions were limited to the consideration and preparation of reports on all waters adjacent to the boundary, from the head of the Great Lakes to the St. Lawrence River and thence to the Atlantic Ocean, with regard to the maintenance of suitable levels, diversions for power purposes, and improvements for navigation.

The story of the International Waterways Commission will be more fully discussed in a later context; here it will suffice to note that it was instrumental in establishing broad principles for the government of boundary waters, which led to the later formation of the epochal International Joint Commission. The IWC itself, however, was an advisory body only, and while it made recommendations concerning the Niagara River, these could not be implemented without international agreement. Pending the negotiation of such agreement, the U.S. Congress in 1906 passed the Burton Act as a stopgap measure. Intended ostensibly for the control and regulation of the waters of the Niagara River and for the preservation of the Falls, the Act actually brought the river under the control of the Secretary of War and empowered him to limit both diversions to American companies actually producing power and the quantities of power to be imported from producing companies on the Canadian side of the river. The authorized diversions were not to exceed 15,600 cubic feet per second and the Canadian imports 160,000-hp. The Act was given a life tenure of three years: upon its termination in 1909 the Secretary was to terminate all permits offered.

Approval of the Burton Act on June 29, 1906, resulted in the immediate cancellation, on the ground of non-fulfillment, of four of the eight franchises already granted on the American side of the river; of the four companies remaining, two were actively generating electricity on a large scale. Temporary import permits, subsequently extended for three years, were given to the three companies on the Canadian side. The Ontario Power Company was permitted to send into the United States 60,000-hp, the Canadian Niagara Power Company 52,500-hp and the Electrical Development Company 46,000-hp. To expedite negotiation of a United States-Canadian Treaty covering joint use of Niagara Power, President Roosevelt was authorized, by the provision of the Burton Act, to make the necessary approaches to the British Government.

Pending the signing of a treaty agreement, the indicated policy for Hydro was obviously to buy power, rather than attempt to generate it. Beck was in an excellent bargaining position with respect to the private companies, two of which—the Canadian Niagara Power Company and the Hamilton Cataract Power, Light and Traction Company with its plant at DeCew Falls—had supplied reports to his Commission regarding power generated and the prices at which it had been sold. The other two—the Electrical Development Company and the Ontario Power Company—were still recalcitrant. When the Park Commission decided to charge them rentals on peak instead of average power consumption, the consequent litigation, which will also be examined in a later context, was prolonged for many years. At this time the three plants on the Canadian side of the Niagara had an installed capacity of 117,000-hp and were selling most of their output to customers on the American side.

In the summer of 1906, of course, the whole decision as to public ownership was still in the discussion stage, despite the existence of enabling legislation in the form of the Hydro Act and the appointment of a governing authority. Municipal ratepayers had yet to vote on the proposition and among them were many still poorly informed upon the subject. In the eyes of adherents of public ownership, Beck had already become another St. George, dedicated to destroying the dragon of private ownership. But in the eyes of those who regarded property rights as sacred he was a Minotaur, half-bull, half-man, bent on the ravishment of potential dividends. In such a situation, victory would go to the side which could marshall and disseminate the most persuasive arguments. In consequence, the last five months of 1906 saw both sides engaged in propaganda campaigns of mounting intensity. Considering the period, it might be assumed that the publicity techniques invented or adopted must have been rather primitive or at best naïve. On the contrary, on examination they prove to be little different from those practiced by Madison Avenue, Bay Street or Philips Square today. Beck particularly showed himself to be a master of psychological warfare—sufficiently so to place him in a unique position among Canadian publicists. By this time he had become fanatically dedicated to the cause of public ownership and passionately determined to make the Commission's plans and policies prevail. The late summer and early fall of 1906, therefore, saw him and his engineers continuously on the move from one Ontario community to another, preaching the

Hydro gospel, and setting forth its promises in factual terms, in preparation for the trial of strength which would come on January 1, 1907, when municipal elections would be held throughout Ontario.

In contrast, the propaganda techniques adopted by the Canadian private electrical interests appear now to have been both ineffectual and uninspired. One example transpired toward the end of the 1906 struggle for public opinion, when a meeting was arranged at the Engineers Club of Toronto at which some of the most prominent men of the profession in Canada issued a manifesto that labelled the engineering features of Beck's program technically impractical and financially impossible. When the engineers' opinions were discounted in the press as those of "interested or disgruntled" parties, an advertisement appeared in the Toronto *Globe*, over their joint signatures, describing the Commission's estimates of costs as fantastic. As if to affirm the *ex parte* nature of this statement, among the signatories was R. E. Black, President of the Canadian Electrical Association and a former chief engineer of the Toronto Electric Light Company.

Beck's reply was made on December 20 at a public meeting held to celebrate the first delivery of power to Toronto the previous month, over the transmission line of the Toronto and Niagara Power Company. At the meeting, Beck predicted immense benefits to the city and province if the vote should favor Hydro. It did so overwhelmingly. By a vote of 11,026 to 2,907 the Toronto City Council was authorized to enter into a contract with Ontario Hydro to supply the city with 15,000-hp or more continuous current at from $14 to $18.10 per hp per annum, this price to include all charges for interest, sinking fund, construction and operating costs, etc. Coincidentally with the Toronto election, eighteen other municipalities voted authority to their city councils to negotiate for membership in Hydro's municipal partnership. They were: Berlin, Brantford, Galt, Guelph, Hamilton, Hespeler, Ingersoll, London, New Hamburg, Paris, Preston, St. Mary's, St. Thomas, Stratford, Toronto Junction, Waterloo, Weston, Woodstock.

Even at this point the battle for public ownership had not been won. Before that could be accomplished, the municipal ratepayers a year later would again be called upon to approve or disapprove the signing of an exclusive contract with Ontario Hydro based on the determination of transmission and distributing costs. The intervening months saw a continuance of the ideological civil war into which the struggle for and against public ownership had now developed.

At the outset in January 1907, the odds were weighted slightly in favor of public ownership when the U.S. Secretary of War, William Howard Taft, employed the powers granted him under the Burton Act to regulate American commercial exploitation of the waters of Niagara. The application of existing companies on the United States side to use water and to import power from Canada were alike approved. But the permits were revocable at the pleasure of the President and in the absence of further legislation would expire on June 29, 1909. No longer would individuals or corporations be allowed to make what use they pleased of Niagara's power merely by ownership of the river banks. Henceforth the river was to be the inalienable property of the two nations through whose territories it flowed.

American private-power interests were now watching the unfolding of the Hydro controversy with growing concern. Thus far they had not had to meet a similar threat, but it was obvious that, to check the spread of this ideological infection across the international boundary, they would be obliged eventually to fight a preventive war against Adam Beck and his cohorts. Meanwhile, it might not be too difficult to weaken the enemy's forces by seducing Beck's technical aides.

This in fact proved to be rather easy. As a member of the Commission, Cecil B. Smith was serving without salary, and as investigator for the earlier commission of inquiry he had accumulated much information of value to the private-power companies.

On February 28 Smith resigned. His subsequent services to the private power interests were of critical importance. It was on the basis of his recommendations that the commercial companies were able to forestall Ontario Hydro by developing power sites in the Rideau district of Eastern Ontario, which Ontario Hydro was subsequently obliged to purchase at an inflated cost. Although Smith's defection involved no violation of professional ethics, Beck regarded his departure as an act of treachery to the public service and to himself.

Smith's successor on the Commission was W. K. McNaught, the manufacturing jeweller who was now a member of the Provincial Parliament and a seasoned public-ownership partisan. Another appointee, carried over from the commission of inquiry, was Clarence Settell, who now became Beck's confidential secretary and the author of a pamphlet entitled *The Genesis*, one of Ontario Hydro's first educational publications.

During this period the preparatory work for the construction of Ontario Hydro's transmission lines from Niagara to the municipalities went forward rapidly under the direction of Chief Engineer Sothman and his assistants. Harry Acres, subsequently Ontario Hydro's chief hydraulic engineer, E. W. Richards and F. T. Stocking performed preliminary surveys along the right of way.

In the Legislature an act was passed giving the Provincial Treasury authority to finance the Commission with a 4 per cent loan, the interest to be a first charge on Ontario Hydro's revenues from the sale of power to the municipalities. The Legislature also enacted, on April 20, 1907, an amendment to the Hydro Act designed to remove the restrictions imposed by the Conmee clause of the Ontario Municipal Act concerning the activities of the municipal councils. The amendment further directed the Commission to provide each applicant municipality with a provisional contract based on the maximum price per horsepower at which electricity would be supplied, and an estimate as to the cost of the delivery of the power, this provisional contract to be submitted to the ratepayers for approval.

The Commission now proceeded to place its first contract with a commercial generating company. By this contract the Ontario Power Company, owned in the United States, agreed to furnish power at 60,000 volts (later raised to 110,000 volts) up to 100,000-hp of three-phase 25-cycle alternating current, on a 24-hour basis. The reasons for the use of 25-cycle frequency will be discussed in a later context: its adoption in this and other contracts was unfortunate, in that Hydro was obliged to continue the 25-cycle system long after the technical superiority of the 60-cycle frequency had been demonstrated and was in general use elsewhere.

Delivery of the contracted power was to be in blocks of 1,000-hp, starting with a minimum of eight such blocks, price at Niagara Falls, $10.40 per hp per annum up to 25,000-hp and for larger amounts $10 per hp. The contract was for a ten-year term, renewable at the Commission's option by ten-year periods until 1950, the Commission reserving its right to expropriate the company's plant at any time, with arbitrated compensation.

From the company's point of view it had done well, since this contract assured the sale of its entire power output. It had been able to sell to the Niagara, Lockport and Ontario Power Company 35,000 of the 60,000-horsepower which the Burton Act permitted it to sell in the United States. Counting the prospective output of its plant, this left a surplus of 180,000-hp, for two-thirds of which Canadian customers would have to be found. Hence the company had every reason to consider its sale to Hydro, whose potential power requirements could be all but unlimited, as a good bargain.

The Electrical Development Company was less eager to trade, having contracts with the Toronto Electric Light Company and the Toronto Street Railway at the high rate of $35 per metered hp or a 24-hour flat rate of $25. In addition to these outlets, the company expected to find a market for its future output in a new industrial community which it planned to establish on a large tract of land already purchased within a few miles of Niagara Falls.

Beck for his part was content to postpone a showdown with Hydro's principal foe until the Commission had its own generating plants. By that time, he hoped, it would be apparent to the Syndicate that the market provided by Hydro's expansion would be sufficient to absorb the Electrical Development Company's output at a fair price and thus safeguard the investors who had put money in this and any other private companies that Hydro might eventually absorb.

Meanwhile Beck's contract with the American-owned Ontario Power Company gave him a bargaining advantage in dealing with its Canadian rival. In May 1907 he informed the directors of the Electrical Development Company that particular territories would be allocated to competing private companies and that these would be made known at a later date, along with the minimum power required, such minimum to be paid for whether taken or not.

Six months later on September 23 the Electrical Development Company was offered its allotment: all the power to be taken by municipalities east of Hamilton, the Ontario Power Company having been allotted the territory west of Hamilton, with a paid-for minimum of 8,000-hp. The Electrical Development Company, however, was to be guaranteed only a 2,000-hp minimum, with the stipulation that "whenever the Commission shall have call for more than 60 per cent of the power then being generated, the Government is to be entitled in case, in the opinion of arbitrators, there is continuous neglect in service, to (claim) forfeit (of) the whole plant of the Company and take it over from the Company."

To these and other proposals restricting distribution, the company's directors replied by letter to the Commission in October setting forth the Electrical Development Company's position:

> The Company has been anxious to meet the Government upon some equitable basis and has earnestly endeavored to secure terms which, while not interfering with the plans of the Government, would leave the Company free to develop and carry out its own business policy.
>
> The Company calls attention to its original franchise from the Government which was ratified by the present government at its Session in 1905, wherein it is provided that, in consideration of the rentals to be paid by the Company to the Government, exceeding the sum of $100,000 per annum, the Government would not compete with the Company in the generation or

sale of Niagara Power, and respectfully insists that in entering into competition with this Company in the way proposed, a breach of the agreement was (is) contemplated.

The franchise from the Government is for the generation of 125,000-hp which the Company is authorized to sell in the Province of Ontario and in the United States. If the Company were to accept the offer made, the following consequences would necessarily result:

1. The Company would be prohibited from selling in any part of the Province of Ontario . . . with the single exception of the City of Toronto. . . . It would have to abandon the whole of its contemplated business in Canada to the Hydro-Electric Commission and be forced to seek a market in the United States for all its power, except that portion which might be taken by the Commission for consumers in the territory allotted to the Company.
2. The Company . . . has purchased a right-of-way to the City of Brantford; this would have to be abandoned.
3. The Company owns 650 acres of land within three miles of the Falls which it has purchased as sites for factories and industrial enterprises; that project would have to be abandoned.
4. The plans of the Company contemplate the sale of power to railways which are to be built in the future and which may traverse the whole territory allotted to the Ontario Power Company. All this future business would be lost to the Company.
5. The Company is already doing business in the territory . . . using only a small proportion of its total output of power, and it is asked to abandon the whole of the rest of its future Canadian business . . . in order to secure the sale of 2,000-hp to the Commission.
6. The Company has expended eight million dollars upon the works at Niagara and its transmission lines, and in order to complete the undertaking further moneys are required; if the Company were to accept the restrictions and limitations sought to be imposed by the Government, it would be impossible to ask from anybody, anywhere, further financial aid.

In the New Year's Day election of 1907, when Toronto and eighteen other municipalities voted to contract with Hydro, Beck had won an important first victory, but obviously the opposition had no disposition to surrender. Instead, the private interests redoubled their efforts to discredit Hydro in preparation for the critical election to be held on New Year's Day 1908.

Even as late as October many voters were uncommitted and victory could still be won by the strongest propaganda. Seldom before in the history of Ontario had the democratic process generated so much heat. Thoroughly alarmed, the private power companies decided to gamble some of the profits they would lose anyway if the money by-laws were passed. Billboards along the highways warned the traveller of the perils of public ownership. Newspaper editorials and, where editorial support was lacking, paid advertisements derided Beck and his supporters as impractical dreamers. Canvassers employed by the Syndicate went from house to house preaching the superior virtues of private power and ridiculing the claims of its opponents.

If Beck had been any of the things his enemies called him he would have been overwhelmed. However, dreaminess and impracticality were not among his distinguishing characteristics. As a strategist he had a clear understanding of where the opposition was vulnerable; as a tactician he displayed both resourcefulness and ruthlessness. The private companies, he told his audiences, were planned and organized to despoil the people of their rightful heritage; their financial structure was devised to collect not one, but a series of outrageous profits. What was in reality one company was organized as three. One generated electricity, a second transmitted it, and a third distributed it, so that three profits might be paid out of the exorbitant rates charged the consumer.

Hydro, by eliminating all these profits, could promise rates for Niagara power which, far from being visionary, were entirely practical. Did the people want cheap electricity or not? Was Toronto prepared to be victimized as Montreal had been by the amalgamation of competing companies and the subsequent imposition of outrageous monopolistic rates for light and power?

As a result of the Montreal merger the private-power monopoly was able to pay 7 per cent on a 24-million dollar capitalization although the actual investment was only $7,000,000. Was that what the private-power barons meant when they talked about the sanctity of private property and wept for the widows and orphans who had invested their all in the companies organized by predatory stock promoters?

With cheerful brutality Beck named names: William Mackenzie, Henry Pellatt, Senator George Cox—members of the Syndicate which was selling Niagara power to Toronto through interlocked companies at notoriously excessive rates. Ontario users, he insisted, should be asked to pay no more than the lowest rate charged for power exported to the United States—$12 per hp, and Hydro had written that stipulation into its contracts with the generating companies. If the squeezed and exploited people of Buffalo had had that kind of protection, he pointed out, they would now be paying $15 or at most $20 instead of $30 per horsepower.

Vote the money by-law, urged Beck, and enable Hydro to end that sort of thing. Galt would then have to pay only $18 per hp, Guelph $19 and Berlin $20. There was nothing unrealistic or visionary about these promised rates. They were what could and would be charged by an honestly administered public authority.

The private-power lobby had money. Beck had people: scores and hundreds of enthusiastic supporters in all the municipalities. Most of them were young: students, engineers, young married housewives, trade unionists, and even members of the Canadian Women's Suffrage Association. They worked as crusaders, explaining and exhorting from door to door.

The private-power lobby too had its volunteer troops, organized under the banner of the Anti-Hydro Citizens Committee of Business Men. Bankruptcy was certain, warned this Committee, if Hydro won. Public debt would mount, municipal tax rates would soar. The very foundations of the Province would be shaken.

Since many of the public-ownership advocates were themselves successful businessmen, like Detweiler and Beck himself, the word "business" had lost something of its magic. Realizing this, the power lobby extracted what propaganda horsepower it could out of another reliable invocation, the "technical expert." Engineers were recruited to deny the accuracy of Hydro estimates and scoff at the notion that power could be brought from Niagara to Toronto and sold at Hydro's promised prices. Even if such rates were practical, they asked, how could inept, inexperienced and notoriously incompetent municipalities be expected to manage such enterprises without courting financial disaster?

As the 1908 New Year's Day election neared, both the news and editorial pages of the provincial press devoted increasing space to the controversy. In a leading editorial in its November 27 issue, the *Toronto World* said:

> The Toronto Electric Light Company are spreading the idea that citizens generally will have to pay the $2,700,000 to be voted on in the by-law for the distribution of power. The astonishing thing is that a number of people believe such statements.

No taxpayer will have to pay one cent for electric power or power plants in Ontario under the plan of the Hydro-Electric Power Commission. The cost of the debentures to be issued will be entirely defrayed by the consumers of power and by them alone. The charges will include interest, sinking fund, depreciation and all other liabilities and will be spread over a term of thirty years. At the end of that time the City will be presented with the plant, free, gratis. No taxpayer will have paid one cent towards it, but the City will gain that asset as a return for lending its credit. . . . The electric ring (the Electrical Development Company which had begun to supply power) will try to convince people that power cannot be as cheaply distributed through a plant which costs $2,700,000 as through their own, which cost $5,000,000 and needs another $1,000,000 now to put it in condition. They will issue pamphlets to show that it is better to pay more to a company that has to pay 8 per cent dividends than less to the City which has no dividends to pay.

By this time the Syndicate had in fact already said most of the things that the *Toronto World* predicted it would say, in its desperate efforts to persuade an increasingly informed and skeptical electorate. Manifestly, it had failed. The votes were still to be counted but the public-ownership forces had every reason to be confident. At a rally of his supporters on December 29, Beck purred like a kitten:

"There was never any intention," said Ontario Hydro's Chairman, "on the part of the Government of Ontario, or on the part of any municipality to do them [the private power companies] injustice or to treat them otherwise than with kindness and consideration."

Hearing this, the beaten and exhausted Syndicate promoters could only pray that they might be delivered from that sort of kindness.

8

In the 1908 New Year's Day elections the public-ownership forces repeated their victory of the year before. Toronto, Berlin, Galt, Guelph, Hespeler, London, New Hamburg, Preston, St. Thomas, St. Mary's, Stratford, Waterloo and Woodstock all voted for the debenture by-laws by large majorities. With the exception of Galt, whose mayor had refused to sign the contract, the thirteen municipalities that had approved the by-law now proceeded to make contracts with Ontario Hydro. Ingersoll was added to the list when the ratepayers reversed their earlier vote against the by-law. Hamilton, Brantford and Paris, which had advantageous contracts with the Cataract Power Company, remained for the time being outside the Hydro circle.

Hydro's first contract with the municipalities was dated May 4, 1908. By its terms the Commission agreed to supply the municipalities with the electricity they needed, and to build the transmission lines and transformer stations from Niagara Falls for the purpose. The municipalities agreed to buy the electricity in specified amounts, and to buy from the Commission only.

At the outset the total amount contracted for by the municipalities was 25,035-hp at a frequency of 25 cycles. The prices charged were to be based on the amount of current bought, and upon the distance to which it had to be transmitted. Toronto took most—10,000-hp—and the price was the lowest, $18.10, the City's liability for capital cost being $828,080; New Hamburg took least—250-hp—and the price was the highest, $29.50, its capital-cost liability being $47,830. Between these two, the requirements of the other municipalities ranged between 300 and 5,000-hp, at prices varying from $22 to $29.50 per hp.

The items which made up the rates to be charged included the contract price paid by the Commission to the Ontario Power Company for current delivered at Niagara Falls; 4 per cent annually on that part of the construction cost undertaken for the municipality concerned; a yearly charge to create a sinking fund sufficient to pay off in thirty years the municipality's part of the construction costs; the municipality's share of the line loss and operating (maintenance) insurance charges. The municipality in turn charged its consumers prices which included the Commission's assessment and the cost of distribution through its distributing system.

With one exception, Ontario Hydro's rates, which averaged $22 per hp yearly, were lower than those charged by private companies in Ontario. The exception was Hamilton, one of the municipalities served by the Cataract Power Company, with current generated at DeCew Falls by water drawn from the Welland Canal.

For eleven years, since 1898, the Cataract Power Company of Hamilton had been generating power at DeCew Falls on the Niagara escarpment and transmitting it to Hamilton, thirty-five miles distant. The storage area was small but the fall over the escarpment at this point was 265 feet. From one penstock and two Swiss turbines with a generating capacity of 3,000-hp, engineering imagination and skill by 1908 had provided an installed capacity of 52,000-hp, an amazing technological accomplishment in view of the contemporary use made of the greatly larger volume of Niagara Falls. So economical was the installation in fact that the company was able not only to supply consumers in Hamilton, Brantford, Grimsby and other communities around the head of Lake Ontario with low-priced current, but to establish a system of electric railways radiating from Hamilton. In all, the Cataract Power Company controlled fourteen public utilities in the Niagara peninsula.

Although it was once estimated that the Cataract Power Company could sell power at $4.50 per hp and still make a profit, it charged Hamilton 10 cents per kilowatt hour for residential current and 15 cents per kilowatt hour for commercial use, in addition to a meter charge. For each street lamp the city was charged $84 per year.

With the coming of Ontario Hydro, the Hamilton City Council at last had a power yardstick with which to bargain. Invited to tender bids, Ontario Hydro estimated $43 per year for arc-lamps, whereupon the private company promptly dropped its rate to $47, but tied the city to a five-year contract at that rate. Later, when the City Council invited tenders for power to pump water and sewage, Ontario Hydro's bid was $17.50 against the private company's $45. Challenged with this disparity, the company then declared that it was prepared to supply Hamilton and other municipalities with current at 10 per cent less than any rate Hydro might choose to offer.

Since it was obvious that in the past, before Hydro appeared on the scene, the company had charged all the traffic would bear, the City Council concluded, not unreasonably, that Hamilton had better safeguard its future by building its own distribution system and buying power from Hydro. Accordingly, the Council voted against renewing the Cataract Power Company's contract when it expired in 1911, and appropriated $505,000 for the construction of a distributing system, to be operated in coöperation with Hydro. The wisdom of this decision was made apparent in 1913 when Hydro's charge for power in Hamilton became $17 per hp per annum. The domestic rate, which the private company had been forced to reduce to 8 cents per kilowatt hour, was soon reduced by Hydro to a little less than one and one-third cents per kilowatt hour, with no meter charge.

In Galt the Mayor had refused to sign the contract with Hydro, on the ground that it differed substantially from that approved by the ratepayers. Now the Mayor was asked to sign a different contract specifying a price of $10.40 per hp per year at Niagara Falls, plus variable charges for other items. In a *mandamus* proceeding, Mr. Justice Anglin found that the contract did not in fact conform to the by-law, that such a contract would therefore be illegal, and that the Mayor "was justified in refusing to become a party to the perpetration of these illegal acts."

Other legal actions aimed at halting the spread of public ownership were initiated by ratepayers in Toronto and in London, who charged that their respective municipalities had acted illegally in contracting with Hydro. Defense counsel successfully objected that the proceedings were not properly constituted, since Hydro was not a party to them and since, under the Hydro Act, the Commission could not be sued without the consent of the Attorney General. Mr. Justice Latchford, who heard the case, then directed that the Attorney General should be asked to join the Commission as defendants. Premier James Whitney, then Acting Attorney General, declined this request until the matter could be considered at the next session of the Legislature.

As was expected, the Legislature acted promptly. In March 1909 it passed the Power Commission Amendment Act. The effect of this amendment was to stay any legal action which had been brought or was pending. It removed Hamilton and Brantford from the list of coöperating municipalities and included Ingersoll and Galt; it required the Mayor of Galt to sign the contract. Thereby it amended the earlier validating act of April 14, 1908, and established the legality of the debenture and contract by-laws, so that money raised by loans or bonds for which the security rested on the by-laws was no longer in jeopardy.

As Hydro entered a period of competitive co-existence with the private companies, the activities of the Commission were housed in a few rooms in downtown Toronto, opposite the Post Office. In 1907 the Commission moved to a suite of offices in the Continental Life Building at the corner of Bay and Richmond Streets; then in 1916 to its newly constructed office-building on University Avenue adjacent to Queen's Park, which became the executive and administrative headquarters of the Commission. An overworked staff strove constantly to keep pace with Beck's drive for expansion, which affected all the departments. For the coöperating municipalities the Commission's engineers had to estimate the cost of building the distribution systems and provide aid and advice on a multitude of engineering and administrative problems. Meanwhile, another team of engineers, surveyors, draftsmen and business managers were busy with the Commission's own construction program.

For almost a year, the Commission's survey parties had been in the field, carrying their equipment with them in horse-drawn wagons, determining the route for the high-tension cables that would carry 100,000-hp of electric current from Niagara Falls to Toronto. In addition to the surveyors there were teams of young men, many of them students at the University of Toronto, whose task was to persuade farmers and other property owners to grant the necessary easements over their land. This was seldom easy. Electricity was still considered by many a dangerous, diabolic force, and Ontario Hydro field men had to combine the canniness of the negotiator with the zeal of the missionary, preaching the social gospel of coöperation and public ownership. Often they addressed meetings in towns and villages, and when elections were held, shepherded the voters to the polls. The right-of-way determined, engineering teams were sent into the field to select the site for the hundreds of sixty-foot steel towers which, to many, in the words of Howells, would be an impertinence to the landscape. Aesthetic prejudice, however, was a relatively minor consideration. What made the preliminary work of the Ontario Hydro crews particularly difficult and costly were the machinations of the Syndicate and its agents.

The newly built line of the Toronto and Niagara Power Company, an affiliate of the Electrical Development Company, was already in operation. It occupied a 100-foot right-of-way across the counties facing on Lake Ontario and provided space for a projected high-speed electric railway between Toronto and Niagara. The line operated at 60,000 volts, considered in 1908 the highest voltage at which current could be economically transmitted. When the Syndicate learned that Ontario Hydro planned a competing line to carry 110,000-volt current, it sent its emissaries into the field with instructions to propagandize the farmers along Hydro's projected route, urge them to boost their prices, warn them of the "danger" of the new, untried high voltage, and incite them to bring damage suits. Later, when Hydro counted up the cost, it estimated that it had been forced to pay about twice as much for its right-of-way as would otherwise have been necessary.

In addition to harassment by the Syndicate, Hydro faced other obstructions. On the outskirts of Toronto, land needed for the transmission line had been expropriated by the Grand Trunk Railway Company. When the company refused to yield its rights, Hydro appealed in vain to the Dominion Railway Commissioners in Ottawa. Later, Ontario Hydro engineers found another feasible entrance for the transmission line along the waterfront; meanwhile, months of time were lost.

The transmission of current at 110,000 volts was one of the technical innovations introduced by Ontario Hydro's youthful engineering team of Gaby and Acres working under Sothman. To cope with the problems of delivering current at the lowest cost possible to the fourteen coöperating municipalities, the engineers refused to accept the current dogma that 60,000 volts was the highest voltage practicable. They recommended that it be increased to 110,000 volts, and the Commission accepted the recommendation.

Other pioneering developments in connection with the use of 110,000 volts were the introduction of the ball-and-socket joint suspension insulators, which eventually superseded the earlier upright pin-type insulators, and adoption of the steel core aluminum cable. Soon these innovations became standard equipment in every country, just as high-voltage transmission, first at 110,000 volts, and ultimately at 220,000 and 230,000 volts, soon replaced the 60,000-volt system used by the Syndicate for its Niagara Falls-Toronto line.

Transmission at 110,000 volts was dangerous and impractical, the Syndicate executives claimed. So were the other technical departures adopted by those hare-brained innovators, Gaby and Acres, unqualified youngsters daring to set the pace for the industry. But Gaby and Acres went ahead and the Commission backed them. Ontario Hydro's transmission line from the Falls to Toronto could be built, they estimated, for $3,500,000, a figure considered absurdly low by contractors employed by the Syndicate. Sir Edmund Walker, with financial interests as head of the Canadian Bank of Commerce, took this occasion to accompany Messrs. Mackenzie and Nicholls in a visit to Premier Whitney. To allow the municipalities to build their own transmission lines was bad enough, they protested, and a subversion of the rights of private property, but to foist this hare-brained project on the public under false pretenses with respect to its cost was intolerable. The Premier, declared Sir Edmund, was being fooled by young whippersnappers just out of college. Did he really believe that the line could be built for $3,500,000? Why, the job couldn't be done for $12,000,000!

After the delegation left, Whitney talked to Beck, and Beck talked to his engineers. They said they would stand by their estimates and so, decided Beck, would he. The Commission Chairman went further: he said he would stake his own reputation and his political future on their professional judgment. With some difficulty Beck reassured the Premier, and the work went ahead. As noted, Hydro had to pay twice what it had estimated for the right of way, thanks to the Syndicate's machinations. Also, to quiet the fears of frightened farmers, the engineers had to build the transmission towers more strongly than was necessary and use twice as many insulators as were actually needed. But Gaby and Acres had wisely embodied wide safety margins in their estimates. The job didn't cost the $12,000,000 predicted by Toronto's leading banker. It cost $3,350,000—$150,000 less than the original estimate. Essentially experimental, its later operation involved the development of improved tower design for the longer spans required and the production of cable able to withstand "aeolian vibration," the tendency of long cable spans to develop failures in light winds. All of these were contributions to the evolving technology of electricity in which Ontario led Canada.

Work on the project was begun on November 18, 1908, at a sod-turning ceremony attended by representatives of all the coöperating municipalities. The first sod was lifted by Alderman Fryer of Galt, President of the Niagara Power Union. On the speakers' platform were Premier Whitney, Commissioners Beck and McNaught and other dignitaries, but Commissioner Hendrie was again absent. The Premier spoke of the obstructions placed in the path of public ownership by its enemies, and Beck triumphantly predicted the victorious advance of Hydro from Windsor to Belleville and on over the whole province, giving the people power at even lower rates than had been promised. The inauguration of Hydro, he declared, would some day rank in historic importance with Confederation.

Early in 1909 the generating companies at the Falls experienced troubles of two kinds: operational and legal. Following a heavy ice jam in the upper river in January which reduced the flow at the Falls to a mere trickle, water levels in February dropped so low that the powerhouse of the International Railway Company ceased generating, and the larger plants on both sides of the river were placed in serious jeopardy. There followed in April a sudden spring thaw releasing the water of Lake Erie and forcing the ice jam to move down river. Below the Falls, in the narrow gorge between the bridge, the jammed ice formed a stupendous dam against which the pent-up water rose until it reached the sills of the Ontario Power Company's powerhouse and then poured over them to inundate the interior with grinding, swirling ice which submerged the floors and platforms and finally filled the generating pits until all was awash like the engine room of a liner sinking at sea.

Fortunately the machinery suffered no serious damage so that by late April, after the floodwaters had subsided, the engineering staff were able to bring the generators back into commission, and the Ontario Power Company could continue the policy of expansion dictated by its growing volume of business with Hydro and other customers. By installing the second of three huge feed pipes under Niagara Falls Park as provided for in the Company's contract, water was obtained for two more 12,000-hp generators, raising the capacity of the plant to 90,000-hp. The other generating companies were also expanding. The Canadian Niagara Power Company had put into operation five units of 10,000-hp and one of 12,500-hp while the Electrical Development Company, with four generators of 10,600-hp each, contracted at this time for three other units of 14,000-hp each. Gradually, Canada was catching up with the United States in its utilization of the Niagara power resources.

The legal troubles encountered by the private companies during this period involved conflicting interpretations of their water rental contracts with the Queen Victoria Niagara Falls Park Commission. Both the Canadian Niagara Power Company and the Electrical Development Company reached the stage, in 1906 and 1907 respectively, where such rent became payable, and the Ontario Power Company found itself in the same position in March 1908. The Park Commissioners insisted that the companies were obligated to pay rentals based on the peak capacities of their generating units, utilizing the amount of water rented to them, whereas the companies argued for rates set at so much per horsepower per year, as determined by the average power generated during the 24-hour day. In February 1908 a conference of Park Commissioners and company officers, with the participation of Attorney General J. J. Foy and the Hon. J. S. Hendrie, MPP, attempted to settle the dispute. No agreement was reached, and the law suits growing out of this conflict continued for years.

During 1908, also, a conference in Ottawa between Canadian and United States representatives discussed the terms of a treaty limiting withdrawals of water from the Niagara River. The treaty, to run five years, was signed in Washington on January 11, 1909, and proclaimed on May 13, 1910. It allotted 36,000 cubic feet of water per second to

Canada and 20,000 c.f.s. to the United States. The Burton Bill, which permitted importation of 160,000-hp from the Canadian side, was not re-enacted by the United States Congress. The Dominion Government at Ottawa continued to regulate the export of power to the United States, through the issuance of revocable permits.

Construction work on Ontario Hydro's transmission line to the Western Ontario municipalities proceeded rapidly during 1909 and the first tower was erected on July 22 of that year. Sites were selected for eleven stations, the step-up transformer station at Niagara, the main interswitching station at Dundas, the interswitching and step-down transformer stations for the municipalities. Contracts for electrical equipment were placed with the Canadian Westinghouse and the Canadian General Electric companies. The Commission's 1909 Annual Report, the first to be issued, gave details of the power contracts it had negotiated—with the Ontario Power Company to supply the Niagara area; with the Kaministiquia Light, Heat and Power Company for the Port Arthur area at the head of Lake Superior; and with the Ottawa and Hull Power Manufacturing Company for the City of Ottawa.

Preparations were already being made for the switching-on ceremonies to be held in 1910. Overworked in every department, sabotaged and harassed from a dozen quarters, expected and unexpected, Hydro's little band of public-ownership pioneers looked forward with mingled hope and trepidation to the approaching moment when the People's Power would for the first time light homes and turn the wheels of Ontario's industry.

9

The passage of the Power Commission Amendment Act in March 1909 closed an important gap in Hydro's defenses. Nevertheless, the attacks of the private-power interests were only briefly checked. Soon the ratepayers' suits against the corporations of Toronto and Hamilton were followed by similar suits against other municipalities, in which various lines of attack were developed by some of Canada's most eminent attorneys.

Sir John Alexander Boyd, Chancellor of the Ontario High Court of Justice, blocked what had seemed a promising legal loophole by deciding that an individual citizen *could* be taxed to pay for a lighting system which he himself did not want, but which the citizens as a whole did want and had installed as a public service. Sir John held that:

> In the present development of economic utilities it may become a question of kind and degree and availableness whether or not the promotion of the interests of a large aggregation of the inhabitants constitutes a public service or not. . . . The transmission and storing and distribution of electrical energy necessitates a system of control and regulation for the interests of public and private safety. . . . The self-interest of the few must give way to the common interests of the whole body of incorporated inhabitants represented by the votes of the majority. . . . The supply of light by means of gas or electricity, with the incidental advantages of heat and motive power connected therewith, appears to be a proper municipal function.

At this point the attack shifted to the Dominion Government, which suddenly received a flood of petitions asking for the disallowance of the Power Commission Amendment Act.

The petitioners must have been aware, despite their claims to the

78

contrary, that even if it could be shown that the provincial power had been abused, the Dominion Parliament had no authority to intervene. All the Constitutional authorities agreed on this point, including a former Dominion Prime Minister, the Conservative Sir John Thompson, and a former Provincial Premier, the Liberal Sir Oliver Mowat. In 1909 the Hon. A. B. Aylesworth, the Federal Minister of Justice, expressed the same opinion.

It was possible, however, for the private-power interests to contend that the Hydro Act was *ultra vires* the constitutional powers of the Provincial Legislature. The petitioners so contended, adding that the Act interfered unduly with vested rights or obligations of contract, and that it did not serve the public interest.

Simultaneously the private-power lobby launched a press campaign aimed at throttling the Province's credit. It took the form of a detailed account of the Hydro legislation and of Hydro's dealings with the Electrical Development Company, written from the Syndicate's point of view. This document was distributed to financial journals, financial writers, investment brokers, and individual investors in Canada and the United States and especially in London, the principal source of capital loans for Canada.

Soon the financial journals of London, Montreal, Toronto and New York blossomed with articles, editorials, and letters-to-the-editor which reflected a common source of inspiration and testified to the widespread nature of the Syndicate's propaganda. Threadneedle Street was shocked. Wall Street was alarmed. Those dreadful Canadians! Misled by demagogues, they had tricked their friends and invited their own financial ruin! The socialistic Hydro legislation was nothing less than government-sponsored robbery, confiscation without compensation! Its intention and effect would be to destroy private companies largely financed by British capital. Did the Canadians realize the inevitable consequences of this sort of thing? Did they wish deliberately to cut off the principal source of the credit on which Canadian industry was dependent for its development?

So went the chorus of bankers, stockbrokers, and financial pundits. In 1909 the private-power lobby gathered the first fruits of its propaganda *démarche* and published them in a booklet entitled: *The Credit of Canada: How it is affected by the Ontario Power Legislation; Views of British Journals and of English and Canadian writers and correspondents.*

The booklet was obviously designed as ammunition for use in the continuing campaign of the private-power lobby to bring about by popular petition the disallowance by the Federal Government, on constitutional grounds, of the Power Commission Amendment Act of 1909.

Citing these petitions, the booklet's introduction noted that "the one main ground which seems to be common to all of the petitions . . . is that by the legislation in question the credit of the Dominion of Canada as a whole is injuriously affected in England, and in the money centres of the world."

The following excerpts are typical of the articles, editorials and letters assembled in this booklet.

In its issue of May 7, 1909, the London (England) *Standard*, after referring to the "apparently unfair" treatment accorded to companies in which British capital was largely invested, declared:

> We should certainly like to know the precise meaning of the schemes which it is affirmed are now being pushed by the City of Toronto for erecting plants in that city in direct opposition to the Toronto Electrical Light Company, also a company backed by British capital.

In an article in the *Financial Times* of London, England, W. R. Lawson, after referring to the clause of the Hydro Acts which stayed court action as "the repeal of Magna Charta," goes on to say:

> . . . such an outrageous parody of law-making must inevitably provoke energetic resistance. How a Conservative Legislature and Scotch to boot [sic], could ever have been cajoled into passing it is a psychological puzzle. The only explanation conceivable is that Ontario is having a very severe fit of municipalizing mania. In order to get a power plant "of the people's own" it is prepared to run any financial risk, to enter into any kind of contract, and to declare legal any kind of illegality. This all means a lovely time for the lawyers, but a bad time for investors, who may by chance get into "validated" Ontario securities. Manifestly the whole question must be thoroughly threshed out and satisfactorily settled before any important Ontario loan, whether Provincial or Municipal, can be offered again in London.

The *Toronto News* published a long series of articles by its editor, J. S. Willison, who declared:

> The proposal of the Ontario government to become a competitor of the Electrical Development Company is indefensible. If it is believed that this company's ownership and operation of the development and transmission plant will be a danger to the

public, let the Government expropriate the plant and thus end the danger. But no loss should be caused to the shareholders or bondholders. In general, public credit has no business in competition with private investment. . . . We are a borrowing country and if we plunder British investors it will be at our own cost. . . . Perhaps for many years every private enterprise in Canada, industrial or otherwise, will find it difficult to obtain money, no matter how promising the enterprise may appear. Sometimes such enterprises will find it impossible so to obtain it. All street or steam railways, all electric lighting companies and other so-called public service corporations will be condemned either to pay excessive rates of insurance against the risk of confiscation or will be obliged to go unassisted. Whether the Municipalities or the Province like it or not they will be driven at all costs to undertake these services for themselves and become constant suitors for money in the very markets from which they have driven private enterprises . . . but they go to these markets with a besmirched character.

The *Financier and Bullionist* published a letter from its editor, Mr. R. J. Barrett, who opined:

Well-wishers of Canada are grieved beyond measure that the British investor has been smitten in the house of a friend—that his reliance on Canadian integrity may seem to him a broken reed, that Canadian credit has suffered in the Homeland's estimation. It is useless to argue that only one instance of seeming repudiation has occurred. The answer is that if the Government of a Province like Ontario violates and rides rough-shod over property rights and declines to play the fair game, then anything may happen in Canada. . . . Consider what the estrangement of British capital from Canada means. At a meeting of the Royal Statistical Society in June a paper was read showing that the total of Britain's Capital investments in other lands amounted to £2,700,000,000; that its increase to £3,000,000,000 by the close of the current year was expected; and that the increase in the seven years from 1905 to 1911 would average £100,000,- 000 per annum. Is Canada disposed to forego participation in that magnificent overflow of fertilizing capital—one that is yearly increasing in volume? Has her limit of expansion been reached? Or is not a very much larger proportion than hitherto of Britain's surplus wealth an absolute necessity if Canada is to fulfil her own great aims and the hopes of her most ardent friends? If so, it scarcely seems wise or kindly to alienate the British investor by treatment he resents.

Elsewhere Mr. Barrett urged the Federal Government to put any possible constitutional curb on the wayward Ontario Government who were driving off needed capital. Canada was warned that British investors—who had assumed that in Canada capital was safe from predatory raiders—would not soon forgive or forget the predatory Hydro legislation.

The Economist declaimed editorially:

> ... Not content with establishing a competing electric supply, the Ontario government took upon itself to forestall or nullify the decision of the Courts of Law by passing an Act which prohibited the bringing of law-suits on this point. What the Canadian Constitutionalists, as distinct from British investors, protest against is the usurpation as they regard it of an absolute power of confiscation and the closing of access to justice by the Government of the day. This Hydro Electric case is the second instance of apparent usurpation . . . other instances of a similar kind are quoted which go to prove . . . that Provincial Legislatures cannot be trusted to deal fairly with the rights of property, and that they have given just cause of alarm to the English capitalist, who is supposed to have fifteen hundred million dollars invested in Canada.

The Economist went on to urge the Liberal Premier of the Dominion, Sir Wilfrid Laurier, to quash the Hydro legislation by declaring Niagara as a navigable waterway, to be under Dominion jurisdiction exclusively.

The Statist wanted to know in what way, after such betrayal of trusting investors, Ontario proposed to find the large amounts of new capital certain to be needed for the further development of the province.

The Financial Post of Toronto asserted that the attempt of the Whitney Government to finance Hydro by raising a three-and-a-half-million-dollar loan had absolutely failed.

In the *Weekly Sun* the 86-year-old Goldwin Smith, who had come from Oxford to America to help in establishing Cornell University, thought that Englishmen, who differ so very much from Canadians in political character, would have behaved differently with respect to the manifest injustice of the Hydro Act. They would not have asked questions about who had been wronged, but would have at once demanded that the injustice be put right.

The silence of the Toronto Board of Trade and of the Toronto press with the exception of the *Financial Post* has been disappointing. To what quarter are we to look for protection of chartered rights against Government violation?

In addition to its excerpts from the financial press, the power lobby's booklet quoted the protests of individuals, contained in speeches and letters. Sir Seymour King, MP, a London banker, wrote the Canadian Minister of Finance that the Ontario power legislation alarmed him. Already, as a result of it, the securities of the companies for developing Niagara Power were at a discount. The credit of Canadian institutions, which had been well thought of, had suffered a severe shock. Eventually, of course, this would be damaging not only to Ontario, but to all the other provinces.

The Professor of English Law at Oxford, Professor Dicey, was quoted as saying that the injustice of the Power Act was almost patent, that the Act was also impolitic, that the petitions against it were justified, and that indeed he could hardly conceive of a stronger case for disallowance.

Lord Ridley, President of the Tariff Reform League, declared that the Ontario Hydro Act overrode the wishes of the ratepayers, made contracts for them without their consent, stayed all actions pending in the courts, and was having a prejudicial effect on the once-trusting but now terribly shocked and wary British investors.

Canada's High Commissioner in London at this time was Lord Strathcona and Mount Royal, then in his ninetieth year. Lord Strathcona became the target for many of the letter-writers, among them the firm of H. Evans Gordon and Company, investment brokers, C. F. K. Mainwaring, an investor in Canadian securities, J. W. Palmer, an executor of the estate of the Duchess of Marlborough, who had invested in the securities of the Syndicate's Toronto Electric Light and Electrical Development Companies, and H. Brent Grotrian, a barrister. Mr. Grotrian wanted the Dominion Government to override the Provincial Government by disallowing the Hydro Act and hoped that Lord Strathcona would make representations in the proper quarter.

At about the same time Mr. Fred W. Field, managing editor of the *Monetary Times* (Canada) said to an interviewer:

> One can quite understand the alarm of the many British holders of bonds of Ontario electrical companies. Such extraordinary and un-British lawmaking is necessarily a blow to Canadian credit in England. . . . Politics should be sidetracked when a nation's financial credit is concerned. . . . One can wink at the

political heroics which are sometimes thought to be necessary
to parliamentary reputation; but it is impossible to overlook and
allow to stand bullying legislation which takes away the first
right of the British subject.

Neither Premier Whitney nor Hydro's Chairman was disposed to
endure these attacks without retort. The private-power interests, said
the Premier, had instigated a campaign of intimidation. They were not
really concerned about Ontario's financial credit. What they expected
to accomplish was to destroy Hydro and bring the Government to heel
by sabotaging the credit of the Province in the loan markets. He was
astonished that reputable British financial papers should accept and
print as true such patently prejudiced statements, all similarly worded,
making serious allegations concerning actions by the Legislature of a
great province. As for the Canadian participants in this clique, "all the
watered-stock experts and stock gamblers in Canada are on the side of
our opponents in this matter and the latter are paying full rates per
line for every word published in their interest by the newspapers in
this country."

Both the Premier and Beck replied to the attack on Hydro with point-
by-point rebuttals of the charges. The Commission, Beck pointed out,
was acting simply as an agent for the municipalities, not directly competing with power companies. The Commission had tried to protect the
business of the Electrical Development Company by offering it exclusive marketing territory, subject only to conditions which would
have prevented it from underselling Hydro outside Toronto, but the
company had rejected the offer.

Continuing the rebuttal, Premier Whitney explained there was nothing improper in the Power Commission Amendment Act passed to
validate municipal by-laws empowering municipalities to enter into a
contract with Hydro. The by-laws themselves had been carried in nearly
all cases by a four-to-one majority, and the Legislature subsequently
authorized the municipalities to sign the amended contract on the
strength of that vote, provided the municipal councils found the amendment acceptable. All the councils voted in favor of the amendment and
all—with the exception of Galt, whose mayor refused to sign—entered
into a contract with Hydro. Although upheld by a court judgment, the
mayor of Galt's decision was later overruled in the Legislature at the
request of the Galt Council, clearing the way for that municipality to
sign the contract. There was no question of illegality in these proceedings insofar as the Ontario Legislature validated irregular by-laws at

almost every session and its power to do so was unchallenged and beneficial.

Concerning the staying of actions at law the Premier, in a statement published in the *Economist*, cited the Education Act recently passed in the British Parliament with a similar intention and effect. It was also pointed out that in the British House of Commons the British Government had submitted as many as thirty-four pieces of legislation which provided that there should be no appeal to the courts in questions directly relating to rights of property.

Concerning Federal disallowance of provincial enactments, the Ontario Government's formal statement quoted a decision of the Privy Council in which Lord Herschell, for the Judicial Council, said:

"The suggestion that the power might be abused so as to amount to a practical confiscation of property does not warrant the imposition by the Courts of any limit upon the absolute power of legislation conferred." The only remedy, he intimated, in the event of an abuse of power, was an appeal to those by whom the Legislature is elected. The statement then added to the pronouncement of the Privy Council the opinion of the Canadian Minister of Justice, the Hon. A. B. Aylesworth. Concerning the notorious Cobalt Lake Mining case, Mr. Aylesworth had said, "It is not intended by the British North America Act that the power of disallowance...shall be exercised for the purpose of annulling Provincial Legislation so long as such legislation is within the power of the Provincial Legislature to enact it."

Replying to the charge that the Hydro Act amounted to a repeal of Magna Charta, the Canadian Minister of Justice said in the House of Commons:

> I am stating to this House that a Provincial Legislature, having as is given to it by the terms of the British North America Act, full and absolute control over property and civil rights within the Province, might, if it saw fit to do so, repeal Magna Charta itself. . . . I take it that no one would dispute the power of a Provincial Legislature to repeal the Habeas Corpus Act or any other charter of liberty which Englishmen possess; and in precisely the same view I take the ground that rights of property are subject only to the control of Provincial Legislatures within Canada.

As to the question of disallowance of the Hydro Acts, the Minister, acting for the cabinet, in April of 1910 dismissed the claim of Federal disallowance, holding that the Act clearly related "to matters declared by the British North America Act to be within the exclusive authority of the Provincial Legislature." He advised the Governor General that he considered it "impossible in accordance with both practice and principle that your Excellency's Government should sit in judgment upon the propriety of this measure." He pronounced the Act *intra vires* but explained that he was expressing no opinion on its merits.

That pronouncement disposed of the agitation for disallowance. The Ontario Government concluded its formal reply to the Application for Disallowance with this ringing declaration:

> Finally, the people of Ontario take their position on the positive and unshaken foundation formed by the British North America Act and respectfully submit that for upwards of two hundred years the Lords and Commons of Great Britain have legislated without fear of the Royal veto although its existence has been undoubted; and therefore, in full accord with the spirit and genius of British institutions, the people of the Province, entitled to all rights of British subjects elsewhere, and as free to legislate within their jurisdiction as the Lords and Commons of Great Britain are free to legislate, cannot submit to any check upon the right of the Legislature to legislate with reference to subjects within its well defined jurisdiction although a technical right to disallow may exist.

Surprisingly perhaps, the inspired attacks against Ontario's Hydro legislation in the English financial press had little or no effect on British investment in Canada. The stock market quotations of the affected companies' shares slumped, but that was all. With the settlement of the prairies in the west, tens of thousands of British and European immigrants were pouring through Montreal and Toronto each succeeding year; the country was in the midst of an unprecedented railroad boom, and capital, grasping the opportunity of certain profits, flowed into the country without abatement. The British investment in Canada actually doubled and redoubled between 1902 and 1914, when it reached £500,000,000.

LIGHTS ON

10

For the small Western Ontario city of Berlin (now Kitchener), October 11, 1910, was a Day of Triumph. There, several years before, the germinal seeds of Hydro had been planted, and there, in the first of a series of "switch-on" ceremonies held throughout the province, was first exposed to public gaze substantial evidence of their fruition—the illumination of streets and public buildings by electricity produced at Niagara Falls one hundred miles away. One hundred miles! One hundred and ten thousand volts! The words held magic as an affirmation of Townsman Detweiler's once-disparaged dream of cheap, publicly owned and distributed electrical energy.

The better to salute the Great Occasion, the City Fathers had seen to it that a part of the main street was newly paved and new street-lighting standards were festooned with as yet lifeless garlands of incandescent lamp bulbs to await the exciting moment. The year 1910 being still in the horse-and-buggy age, the town's numerous livery stables were full to overflowing with teams from the surrounding countryside, but the town's automobile owners, perhaps two score or so at the outside, were also out in force, most of them to chug their way to the Grand Trunk Railway depot where Mayor Hahn, attended by local notables, awaited the special train from Toronto bearing Premier Whitney, most of his Cabinet, several members of the Provincial Legislature, other dignitaries, and the press. Ontario Hydro's Chairman and representatives of other towns and cities of the public-ownership fraternity were arriving by other trains.

Following the ritual of welcome on the depot's pine-plank platform, the official party made its way by motor cavalcade along streets bedecked with banners and lined with cheering crowds to the largest auditorium available: the hockey rink, where a temporary speaker's platform had been erected. There Mayor Hahn, Premier Whitney, Adam Beck, and others, including the local member of the Dominion Parlia-

87

ment, William Lyon Mackenzie King, took their seats. Neither John S. Hendrie nor the Hon. Frank Cochrane, a member of the Whitney Cabinet, dignified the gathering by their presence.

As the Great Moment neared, a little girl climbed the steps leading to the platform, shyly curtseyed, and held out to the waiting Premier a cushion bearing an electric switch. Turning to the Ontario Hydro Chairman, the Premier grasped the latter's hand and guided it to press the switch and fill with light and roaring cheers the rink and streets outside.

When the cheering subsided, Beck paid tribute to the pioneers of Berlin and Waterloo County who had organized the first meeting for the public ownership of power, in 1902, eight years previously, and again pledged himself to carry on the fight that they had started until "the poorest workingman will have electricity in his home."

Premier Whitney's speech was a paean of victory. "We have been attacked, vilified, and slandered," he said; "Large sums of money have been expended in creating and fomenting prejudice and ill-feeling against us. And still larger sums have been expended in conducting a campaign against us outside of Ontario. Our opponents have left nothing undone that could be done, and men and influences from the humblest man in the land up to the Prime Minister of Great Britain were approached in an endeavor to destroy our power legislation and render it impossible for this wonderful new force to be used and enjoyed by the people except on the terms laid down by private corporations and individuals. Further, we have been told it would destroy the credit of Ontario and indeed of Canada. Opponents tried to have our legislation disallowed at Ottawa. . . ."

Then, referring to Adam Beck, the Premier said, "We, his colleagues, can never forget his steady confidence in the result and the bravery and pluck with which he stood up against all attacks."

That evening at the banquet in the Market House the honor of proposing the toast to Hydro fell to D. B. Detweiler, Ontario's earliest and most dedicated crusader for public ownership. As Beck rose to respond, the band played "See the Conquering Hero Comes," the guests cheered, and Beck had great trouble controlling his emotions. In a speech designed to fortify Hydro's friends, answer its critics, and conciliate its enemies, the Chairman touched upon a variety of subjects, among them several charges that had been laid against him and his colleagues. Referring to political patronage, he said, "There has yet to be made by the Commission an appointment because of the political leanings of any one of our officers. Some of our loyalest supporters have been Liberals. . . ." Excoriating the local jealousies then rampant in Ontario, he sprang to the defense of the provincial capital. "Toronto," he said, "has been called Hogtown. I wish all cities in the universe had as much public spirit as Toronto has. She is our biggest consumer. She has made this scheme possible. She is in the battlefield today with strong opposition"—to the enemy, presumably. Touching on still another often-repeated charge, Beck proclaimed, "I will not stand for this talk of mayors and aldermen being crooked grafters."

Continuing, the Chairman praised the press of the Province for putting aside party politics to support the People's Project, and assured the private-power interests that their output would be needed to meet the growing public consumption. There was room for all, he said.

There was indeed a great potential for increased electric service, though not, as it developed, for the co-existence in Ontario of publicly and privately owned light and power systems. Even before Ontario Hydro's transmission of power from Niagara, private companies displayed their awareness of the potential markets by buying riparian rights to power sites and setting up hydro-electric stations to support them. Such purchases were made possible on all streams in Ontario other than navigable boundary waters by their earlier acquisition for lumbering purposes.

For example, the Wanapitei Power Company built the Coniston Generating Station on the Wanapitei River in 1905 to service the Sudbury region; the Simcoe Light and Power Company developed the Big Chute Generating Station on the Severn River, three miles from Lake Couchiching, in 1909; and in 1910 the Seymour Generating Station on the Trent River was constructed by the Electric Power Company to serve the Bay of Quinte district. The Electric Power Company was a creation of the Mackenzie interests brought about by a merger of the Auburn Power Company, the Central Ontario Power Company, the City Gas Company of Oshawa, the Electric Light Company of Oshawa, the Cobourg Utilities Corporation, the Nipissing Power Company, the Northumberland-Durham Power Company, the Peterborough Radial Railway Company, the Peterborough Light and Power Company, the Sidney Electrical Power Company, the Seymour Power and Electric Company, and the Trenton Electric and Water Company, among others.

In addition to the Seymour GS mentioned above, the Electric Power Company by 1911 had in operation the Sidney GS on the Trent, the Auburn GS on its tributary, the Otonabee GS (Peterborough), and the Nipissing GS on the South River (North Bay). All of these were relatively small installations of only a few hundred horsepower serving restricted areas, but they indicate a pattern of obstinate competition which was costly and time-destroying to Hydro's progress.

On the other hand, the switching-on ceremony at Berlin was followed by announcements from other municipalities in Western Ontario that they wished to take their full allotment of Hydro power. In quick succession the Berlin ceremony was repeated elsewhere under similar gala circumstances. Guelph and Waterloo came first, both receiving power on November 13, 1910, while Preston, Woodstock, London, Hamilton, and Stratford celebrated the event before the year was out. Dundas and Hespeler were supplied in January 1911; New Hamburg and St. Thomas in February; Galt and Toronto in March; and Ingersoll and St. Mary's in April.

At the switching-on ceremony in Hamilton, home town of J. S. Hendrie, Commissioner Hendrie was present almost perforce. He turned on the switch, and then said complimentary things about his fellow-Commissioners, Beck and McNaught. Beck for his part used the occasion to say uncomplimentary things about the Cataract Power Company, in which many of Hendrie's friends were shareholders. This company, said Beck, was charging 8½ cents per kilowatt hour for energy brought from DeCew Falls, compared with Hydro's charge of 4½ cents in London for energy brought three times as far.

A few months later Hamilton's ratepayers ended their contract with the Cataract Company and voted $500,000 for the construction of their own distributing system. In 1911 Hamilton signed a new contract with Hydro and within another year the system was in operation.

Beck, who attended most of these inaugural ceremonies, was there to press the switch in person when in November 1910 power was brought to his home city, London. Hydro power came officially to Toronto on May 2, 1911. The elaborate switching-on ceremonies brought delegates from all the other coöperating municipalities and attracted crowds so large they could not be effectively controlled. In front of the Toronto City Hall milling thousands shouted and trampled each other; children were crushed breathless and women fainted. Unable to make himself heard, Premier Whitney cut short his speech and once again guided Adam Beck's finger to the ceremonial switch, but the effect was unexpected. Not only did the lights go on with incandescent splendor; the Premier and other platform dignitaries were drenched with water sprayed by a miniature Niagara Falls above their heads which the impresarios of the ceremony had designed as their *pièce de résistance*.

While an emergency first aid station treated wilted women and wailing children, the officials retreated damply inside the City Hall. There Premier Whitney was able to finish his address.

In April and May of 1911 the last switching-on ceremonies brought to an end the ten-year ordeal of Hydro's difficult birth. The laws had been passed, the private-power lobby had been out-argued and outvoted, Ontario Hydro had built the transmission line from Niagara, and the municipalities had built their distributing plants. But the enemies of the People's Power were neither routed nor silenced. Another round was coming up. Now Beck and his fellow-coöperators had to show that they could make the People's Power work—and at the "ridiculously" low cost promised.

Within six months after the lights went on in Toronto, the demand for power, whether Hydro or private, had quadrupled, and by October 1913 the number of municipal customers in the Niagara System had risen from fourteen to thirty-eight and such new special customers as the Hamilton Asylum and the Ontario Agricultural College had been added; other municipalities had joined the Port Arthur, Severn, and St. Lawrence Systems. That same year, 1913, Ontario Hydro began work on its first producing unit at Wasdell Falls on the Severn River and in 1914 purchased the generating station at Big Chute on the same river, built by the Simcoe Light and Power Company in 1909. Wasdell Falls was Hydro's first hydraulic development; Big Chute, its first owned and operating generating station. The initial output of the former was 750 kw, of the second 4,300 kw.

The principal factors in this rapid growth were the extraordinary demands of Ontario's industries for Hydro's cheap power, the tireless evangelism of the Commission's Chairman, and the brilliant response of Hydro's engineers to the requirements they were asked to meet.

Those demands were not wholly technical of course, but were encountered in every Hydro activity. This was inevitable in a relatively young organization which had to meet simultaneously the objectives of its own socio-economic crusade; the assaults of ruthless and not too ethical competition; the requirements of new municipalities as they joined the system; the evolution of its statutory sanctions; and continuous changes in both its own administrative practices and those of its principal customers, the municipalities, governed by various amendments to the Power Act.

To cite an example of the last of these involvements, under Section 16 of an Act respecting the City of Toronto, assented to March 24, 1911, management of the local power system was placed in the hands of a Commission consisting of the mayor, *ex officio*, and two Commissioners, one appointed by the City Council and one by Ontario Hydro. As will be seen in a later context, the situation thus created, giving Ontario Hydro minority representation, provoked a prolonged and heated political controversy between two appointed bodies, both of which were devoted to the public interest.

The members of the first Commission to serve under the Act of 1911 respecting the City of Toronto were P. W. Ellis, appointed by the City Council, H. L. Drayton, appointed by Ontario Hydro, and Mayor G. R. Geary, *ex officio*. E. M. Ashworth served Toronto Hydro first as Secretary, and later as General Manager. In 1915 the provisions of the Act of 1911 were extended to include all cities of more than 100,000 population. Smaller cities were required by the Public Utilities Act to elect Hydro commissions, only village and township councils being permitted direct management of their own systems. One of these townships, York, eventually exceeded the maximum population figure— then 60,000 as provided for by an amendment made in 1927—but continued the direct management of its system.

As almost from the beginning Adam Beck was Hydro's chief spokesman, promoter and salesman, he was constantly journeying throughout Southern Ontario to direct the Commission's developments and expound the advantages of Hydro to interested communities. In the spring of 1911, while Toronto still awaited the advent of Hydro power, the people of Brockville, one of the attractive towns on the shore of the St. Lawrence River and a member of the newly formed Union of Eastern Ontario Municipalities, invited Ontario Hydro's Chairman to tell a public meeting what the Commission could do for the town. The local consensus was that it could not do much, at least not immediately, because Niagara was too far away. Local companies were charging $45 to $50 per horsepower per year for steam-generated power while in the Trent River Valley the Electric Power Company, mentioned earlier, enjoyed a monopoly position with rates in keeping. Hydro would like to serve Brockville and other nearby communities, Beck said, and might soon be in a position to do so, now that they were organized in the Union of Eastern Ontario Municipalities. The Union had asked for 5,000-hp but if its members took 50 per cent more the rate would be correspondingly less; much less if they took double the amount. He explained that a private company had leased water rights at High Falls on the Madawaska River, one of the largest tributaries of the Ottawa, and had agreed to develop hydro-electric power at this site and deliver it to Hydro within ten months. Another offer had been received from the New York and Ontario Power Company at Waddington, N.Y., and Hydro had signed a contract with this company. Both offers were open for eighteen months.

Waddington power would cost $13 per horsepower for 2,000-hp delivered at Morrisburg; for 4,000-hp, $12.50; for 6,000-hp, $12; for 10,000-hp, $11; for 15,000-hp and upwards, $10.50 per horsepower. Similar power in New York State cost $18 per horsepower. It would cost $652,000 to install the transmission system.

Beck then proceeded to spell out just how the wholesale rate would work out for each of the municipalities comprising the Union, answering skeptics with facts and figures, and leaving Brockville a new convert to the gospel of public ownership.

The outcome of Beck's Brockville visit was not entirely typical. In the early fall of the same year, 1911, he tried unsuccessfully to crack the monopoly that controlled the hydro-electric resources of the Trent River Valley. The Commission had taken an option to purchase, on behalf of the town of Lindsay, the Lindsay Light, Heat and Power Company: the Provincial Government had promised to advance $30,000 of the $230,000 purchase price. Nevertheless, the people of Lindsay and the surrounding district voted against Hydro, and the municipality remained outside the system's orbit until 1916.

During this critical period, while Hydro was struggling to expand, Beck's time and seemingly inexhaustible energies were expended prodigally wherever there was a chance of adding another community to the Hydro partnership. Could Hydro get power to Beachville, a tiny community near Woodstock, asked the village leaders? Yes, answered Beck, using all his eloquence to win over the farmers who gathered to hear him in a little oil-lit meeting room.

Windsor, opposite Detroit and over a hundred miles beyond the termination of Hydro's transmission line at St. Thomas, had been willing to take Hydro power if the municipality were permitted to export some of its purchased power across the international boundary to Michigan. This, it was argued, would help to reduce not only the Windsor rates but also those paid by the other Hydro partners since the municipality would derive profits from its export of power.

The issue caused a temporary split in the Commission. Beck was for anything that would expand Hydro's power load and thus reduce rates, while pointing out at the same time that the export permit must be revocable, to meet Ontario's demands. Commissioner McNaught considered the possibility of eventual cancellation to be a decisive argument against the proposal and opposed it so vigorously as to offend the Chairman. In Beck's home town of London, representatives of thirty-four municipalities met and voted unanimously in favor of Windsor's proposition. In Toronto, however, Controller Spence agreed with McNaught that Windsor should not be permitted to retain the profits derived from power export. He pointed out, too, that the U.S. Burton Bill of 1906 allowed the Ontario Power Company, which supplied Hydro, 60,000-hp of the limited exports permitted by the Act.

The plan was abandoned, but Windsor decided to join the Hydro partnership anyway. In March 1913 the Commission applied for authority to extend its transmission line from St. Thomas to Windsor, and a month later an Order-in-Council granted this authority. The line was completed in August of 1914, extending Ontario Hydro's high-tension double-circuit transmission line from 171 miles to 274 miles.

The tripling of the number of Hydro's municipal partners and the quadrupling of the demand for power imposed mounting burdens on Ontario Hydro's young engineers, by 1913 grouped in four divisions: the Niagara, the Port Arthur, the Severn and the St. Lawrence. Each of these divisions demanded expansion and improvement, requiring the installation of new equipment. In 1913 the Severn System was enlarged by the addition of Barrie, Collingwood and Coldwater, and later, Elmvale and Stayner, making a total of seven communities served. In the Niagara System three separate circuits were installed between Dundas and London, and the capacity was raised at the Niagara, Dundas, London, St. Thomas, Stratford, Guelph, Berlin, St. Mary's and Toronto transformer stations. The engineers were also obliged to design new distribution stations and feeder lines to supply Hydro's steadily growing list of municipal partners.

At Port Arthur, in Northern Ontario, Hydro was supplying a new government grain elevator; also the Port Arthur Elevator Company, which operated the Canadian National Railroad elevator. To supply the St. Lawrence System the Commission had been able to buy current from the Rapids Power Company at Morrisburg. To supply the Severn System, now serving seven municipalities, the Commission undertook the construction of the Wasdell Falls GS on the Severn River, beginning work in July, 1913, and completing it fifteen months later. The Wasdell Falls GS constituted both a school and a proving ground for Ontario Hydro engineers, who displayed great ability in dealing with unprecedented problems.

The year 1913 also saw the completion of a new building for Ontario Hydro's experimental and testing laboratories, which had previously been housed in the basement of the transformer station on Strachan Avenue, Toronto. The original equipment had consisted of a lathe and a drilling machine to make emergency repairs on certain apparatus, together with a small testing bench for comparing the performance of different types of insulators, incandescent lamps, and recording and indicating meters.

In charge was Ontario Hydro's first research director, H. D. G. Crerar, afterward General Crerar, the distinguished soldier who succeeded General McNaughton in command of the Canadian First Army in World War II. A graduate of the Royal Military College at Kingston in the class of 1909, Crerar's first job was with the Canada Tungsten

Lamp Company in Hamilton. In 1911 Adam Beck invited him to fill a position described as "illumination engineer." Crerar started work with a basement bench as his sole equipment and with only a single assistant, P. A. Borden. A year later the laboratory staff was enlarged by the addition of R. B. Young, a graduate of the University of Toronto, who was to serve the Commission for forty-one years and become one of the world's acknowledged experts on concrete construction.

During the same year that saw the birth of Ontario Hydro's research laboratory, the Commission set up an Inspection Department under the direction of H. F. Strickland. This was another of Ontario Hydro's technical and administrative "firsts": the private companies never having bothered to supervise wiring and other electrical installations, poorly trained electricians had remained free to wire factories as well as public and private buildings.

Struck by the pressing dangers of this situation, the Commission determined to establish standards of materials and workmanship and enforce them by legal sanctions. By agreement with the Legislature, Ontario Hydro's engineers undertook the drafting of a code of Rules and Regulations as required by the Power Commission Amendment Act of 1912 which empowered the Commission to enforce the code and require municipalities to appoint inspectors to assure its observance. Armed with this legislation, Ontario Hydro was now in a position to ensure compliance with the code by all its present and prospective municipal partners. Ontario is believed to have been the first province or state in North America to adopt such legislation.

In certain cases this meant a complete replacement of the primitive and dangerous installations commonly in use: drop lights without wall-switch controls, wooden fuse boxes and fuse blocks, open link fuses and switches, and faulty equipment for street lighting. In 1913 H. F. Strickland, chief of the Inspection Department, reported that these and other unacceptable conditions existed in several municipalities. Some objected to the stringency of the Commission's regulations, but all were obliged to conform if they were to continue to receive Hydro power.

At first the Commission gave its approval to inspectors appointed by the urban councils. These inspectors, however, had no jurisdiction in the suburban communities clustered around such cities as Toronto, Hamilton, London and Windsor, hence inspection of electrical wiring remained incomplete until further legislation was enacted by the Provincial Government giving the Commission authority to appoint its own inspectors. The Commission's inspectors were given authority to inspect electric wiring in all premises served with electric power, not only by Ontario Hydro and the municipal Hydro Commissions and utilities but also by installations of private companies. Thus, from 1914 on, power consumers in Ontario enjoyed the protection of legal standards and inspection services which were only adopted in other parts of the world at a much later date.

11

In 1911 Ontario Hydro experienced a threat to its administrative autonomy and in 1912 its first unpleasant scandal.

During the summer of 1911 Adam Beck, the Chairman, and P. W. Sothman, the Chief Engineer, travelled through Europe inspecting hydro-electric installations and equipment. Included in the party were the Chairman's wife, his cousin, Alfred Clare, and W. Bert Roadhouse, Secretary to the Ontario Minister of Agriculture, who was to report on the uses of electricity on European farms.

At some point on the trip, Beck and Sothman quarrelled, for reasons which were never disclosed. When they returned to Toronto they were no longer on speaking terms and for months they avoided each other, while in the Commission's offices gossip spread.

Beck, who believed Sothman guilty of dishonesty, discussed the matter with W. W. Pope, Secretary of the Commission from October 1909, who produced evidence tending to support the Chairman's suspicions: Sothman had several bank accounts and was living beyond his means. Recently he had spent $25,000 on the purchase of some real estate.

Detectives were employed. They obtained evidence that Sothman had exacted rake-offs on contracts. What looked like a bribe of $18,000 had been paid to Sothman by an American supplier of electrical apparatus, the sale of which to Ontario Hydro had been expedited by Sothman.

In July 1912 Beck brought the Chief Engineer before a meeting of the full Commission and presented some of this evidence. It was enough, said the Chairman, to warrant laying criminal charges against him. But before charges could be brought, Sothman fled, crossing the border into the United States. Subsequently he tried to collect arrears of salary from Ontario Hydro, but he never returned to Canadian soil.

Beck had been fighting a political battle to maintain control of the

Commission's function of electrical inspection. His opponents within the Whitney Cabinet, who had been strong from the beginning, were pressing the Premier to transfer control to the Ontario Railways and Municipal Board. They succeeded to the extent of persuading the Premier to draft a bill which embodied the change demanded, and which was introduced in the Legislature on January 25, 1911. The Bill got a first reading, but met unexpectedly strong opposition from Beck supporters in and out of the Assembly. When it came up for second reading, the Premier asked that it be allowed to stand on the Order Paper indefinitely.

"I am not disposed to press the Bill in its present form," said Mr. Whitney. "Experience has shown the necessity of several added provisions."

Political observers suggested that the Premier was being somewhat less than frank; that his real purpose in pigeonholing the Bill was to hold it like a Damoclean sword over the head of the obstreperous Ontario Hydro Chairman, who had repeatedly shown his unwillingness to act as a biddable servant of the Cabinet. Beck understood this, but treated the sword as though it were a papier mâché ornament.

It was under the shadow of this threat that Beck left for his European tour with Sothman to visit London, Munich, Danzig, Baden and Milan. The tides of Canadian politics, he had reason to believe, were running in his favor. That year the Federal general election rejected Sir Wilfrid Laurier's Liberal Party, with its policy of Canadian-United States reciprocity, and swept into office a Conservative Government under Premier Borden. In Ontario, Premier Whitney seized the opportunity to hold a Provincial election; the Liberals were at the moment busy rejecting their leader, A. G. McKay, who had supported the private-power interests on the ground that Hydro had benefited the Niagara district at the expense of the rest of the Province.

The future status of Hydro became an election issue when Whitney, in an address to the electors, declared that:

> In our opinion the time has come when, having regard to the conduct of public business under our system, the Hydro-Electric Power Commission should be discontinued and a new Department of Power created which could take charge of this great work, and the head of which should be a cabinet minister.

Hence, when the election results gave the Conservatives 83 seats to 22 for the Liberals, the Conservative Premier had what he could reasonably have interpreted as a popular mandate to abolish the Commission. Instead, it would seem that Whitney gave greater political weight to the pro-Hydro Liberal Opposition under its new leader, N. W. Rowell, and to the numerous pro-Hydro Conservatives in the municipal governments. At any rate, nothing more was heard of the Bill to put Ontario Hydro under a department of power, and on January 11, 1912, Beck announced publicly that Ontario Hydro would *not* become a government department—not, at least until it was serving areas of the province much wider than the Niagara district. The qualification helped to save face for the Premier, but in effect it meant nothing. Ontario Hydro was now securely established as a quasi-autonomous utility corporation, and everything that happened subsequently tended to confirm this status.

In the 1912 New Year's Day elections, twenty-nine additional municipalities approved the Hydro by-laws, and at the same time, the Toronto ratepayers voted by an overwhelming majority an additional $2,200,000 to complete their Hydro distribution plant.

While Beck's tireless speaking activities contributed greatly to these victories, his political support thus far had been unorganized—too unorganized, in the opinion of many of his followers. To correct the weakness they undertook, in February 1912, to create the Ontario Municipal Electric Association, thereafter to be known as OMEA. Paradoxically, one of the purposes of the new Association was to make Hydro independent of political interference and protect it from the inefficiency of political appointments, then not uncommon in Ontario. Today we would call OMEA a pressure group. Its members were the appointed delegates of the councils in Hydro municipalities. Each council had one vote and paid dues proportioned to the size of the municipality: $10 for the larger communities and $5 for the smaller ones.

The OMEA constituted a powerful group, outside both political parties, yet capable of such political influence that it was, in respect to its own affairs, an arbiter between them. The enemies of Beck seized on this and represented the Association, in conjunction with the newspapers favorable to Hydro, as an anti-democratic machine at the personal service of a dictator. Beck did indeed make full use of the machine, but that was precisely as the municipal appointees wished it. The Chairman for the first year was G. R. Geary, who was followed by J. W. Lyon of Guelph. The secretary was E. M. Ashworth, who later became the General Manager of the Toronto Hydro-Electric System.

The objects of the OMEA as outlined at the first general meeting were:

a: To take united action on all Hydro matters.
b: To unite, as far as practicable, in the purchase of electrical sup-

plies, and as far as possible to endeavor to obtain standardization
of equipment, accounts, operations and general management of
municipal plants.

c: To work in conjunction with the Hydro-Electric Power Commission of Ontario in promoting electrical development in the Province.

d: To suggest such legislation as might be deemed of advantage to
the hydro-electric undertaking of the Province and to take united
action thereupon.

e: Generally to promote the interests of the municipal electric undertaking of the Province.

According to Ashworth, some of the resolutions passed at meetings
of the Association carried the initials of Adam Beck. That, however,
was probably not true of a resolution passed unanimously at the first
meeting, which praised the efforts of Ontario Hydro's Chairman and
asked that "a fair remuneration should be paid for his work of the past
years and provision made for an adequate annual salary as Chairman
of the Hydro-Electric Commission of the Province."

Publicly, Beck had declared himself opposed to the payment of
salaries to the Hydro Commissioners. Privately, however, he acknowledged that his almost exclusive preoccupation with Hydro's affairs had
entailed some monetary sacrifice. Hence it may be assumed that he was
not wholly reluctant to be overruled by the Provincial Legislature,
which at its 1912 session responded to OMEA's appeal by passing legislation which provided an annual salary of $6,000 for the Chairman of
the Commission, in addition to what he received as a member of the
Legislature.

Ascending on the escalator of the electrical revolution, Hydro was
fulfilling the prophecies of its most enthusiastic advocates. At OMEA's
first meeting, Beck was able to boast that the Commission had contracted to supply 44,000-hp, and that the spread of benefits from cheap
and reliable power was proceeding at a rate beyond his wildest expectations. Nine cities, fourteen towns and six villages were members of the
Hydro partnership and agreements were pending with another thirty
municipalities.

Preaching the new gospel of industrial decentralization, Beck rushed
from Parliament to city platform, and from town hall to rural village.
"What we want," urged Ontario Hydro's drum-beating Chairman, "is
that every small village be an industrial centre; small towns and villages
with plenty of manufacturing, rather than great factory centres with
slums and congested populations."

On July 31, 1912, another Hydro pressure group was formed: the Midland Association of the Hydro Power Union. Addressing its members, Beck explained with facts and figures what electrification could and would do for Ontario's farmers. Uxbridge and eight surrounding communities north of Toronto, declared Beck, were now within practicable range of Hydro's service.

In October Beck was in Brantford, which had voted in 1909 for the Hydro by-laws but had continued its contract with the Cataract Power Company. Now, with the help of Mayor Geary of Toronto, where Hydro had reduced power rates by a half-million dollars a year, Beck was able to persuade the Brantford ratepayers to join the Hydro partnership. Paris, Brantford's neighbor in the Grand Valley, together with Port Dalhousie on Lake Ontario, and the municipalities of Brockville and Prescott in the east, also signed Hydro contracts.

At the Brantford meeting, Beck advocated what was later to become an unfortunate obsession: the construction of publicly owned radials (interurban electric railways) as part of Hydro's expanding empire. At a meeting in Toronto a few weeks later OMEA took up the cry, when delegates from sixty municipalities voted unanimously for a resolution which declared:

> That in the opinion of this meeting it is desirable that a system of electric railways, including street railways, to be owned by the municipalities, be established and built; and further that the Hydro-Electric Power Commission be requested to look into the advisability and practicability of building such a system and to furnish a report thereon to this Association.

If the enthusiasm for electric suburban railways seems incredible today, it must be remembered that the condition of Ontario roads was then generally abominable. No provincial highway authority existed, jurisdiction resting with the county councils, where there was settlement, or with the Colonization Roads Division, a branch of the Lands and Forests Department, where there was not. A narrow, concrete road had been built from Toronto to Hamilton but otherwise the roads were macadam where road metal was abundant; elsewhere they were rutted clay, sand or rock.

The cry for the radials, therefore, was not as preposterous as it may seem today. In 1913, in fact, they seemed to be the only means by which improved communications and transportation within the far-flung province could be economically provided. Beck declared that Ontario Hydro would be glad to estimate costs, but financing the Provincial radials would be up to the Provincial Government; it in turn should ask the Federal Government "for the usual subsidy to build a railway."

In 1913 Beck's drive to expand collided head-on with the Mackenzie interests, which had proposed a tempting plan to supply power to the communities of Aurora, Richmond Hill and Brantford, north of Toronto. The project envisaged the construction, by Mackenzie's Metropolitan Railway, which operated the radial north on Yonge Street from Toronto, of a 12,000-volt line to carry current from Toronto to Newmarket, where for months the struggle between Hydro partisans and the Mackenzie interests had embroiled the town and neighborhood.

Mayor Cane led the Mackenzie faction; Councillor Hunter and Colonel W. J. Allen supported Hydro. Voting on May 30, the ratepayers rejected the Mackenzie plan, whereat the Mayor and the pro-Mackenzie councillors resigned. Subsequently Colonel W. J. Allen was elected Mayor by a huge majority, along with a pro-Hydro council. But the Mackenzie faction refused to concede defeat. In the January 1915 election they staged a comeback in which the Hydro by-laws were rejected by a vote of 396 to 287. A month later a by-law was passed by a majority of 178, committing the town to purchase power from the Mackenzie Company.

During these years, the tide of battle between Hydro and the private power interests swung forward and backward over the changing landscape of a province which had now become the heart and centre of Canada's industrial expansion. The private companies fought hard for survival but their rate reductions were too little and too late, and the long-term odds were against them. Hydro was developing and perfecting its administrative techniques and consolidating the victories won by the Chairman's indefatigable salesmanship. Apparatus were standardized, rates unified and standard systems of accounting adopted. The municipal partners could now call on the services of an auditor employed by the Commission.

Between 1910 and 1912 Ontario Hydro's engineering department conducted some thirty-six surveys of the hydro-electric sites in all parts of the province. New recruits to Hydro's municipal partnership required advance estimates of installations and power costs. In 1911 the Commission was faced with the problem of costing each Hydro municipality for the year's service in advance. Concerning the unprecedented difficulties of the problem, A. H. McBride, who served Ontario Hydro's engineering department from 1906 until his retirement in 1949, has written:

Determining the cost of power for electrical service is like no other type of accounting in the world. At the beginning we knew practically nothing about it. Our duty was to supply power at cost, but all the elements involved in that cost were not always an easy matter to determine, and until they were determined it was impossible to tell whether our estimates for power service were away too high or away too low. In either case, we should have been in difficulties, especially as we grew and expanded.

We wrote to the public utilities engaged in the power supply business in England, the United States and Germany, gathering all the data we could on electrical accounting, but we soon discovered that we should have to evolve a system of our own. The trouble was that every cent the Commission spent had to be balanced equitably among the different municipalities. Nobody anywhere, as far as we could discover, had done *that* kind of accounting.

Advance costing was difficult, but so was rate-fixing. To solve this problem the Commission developed another Hydro innovation: the promotional rate structure. As described by Thomas H. Hogg, one of Ontario Hydro's great engineers, this

> consisted of a small service charge [afterwards, so far as domestic service is concerned, abolished in most municipalities], a first energy charge, and a much lower second or "follow-up" energy charge which applied to all power taken in excess of a specified minimum. . . .

Writing in 1941, Hogg says:

> Today the promotional rate structure is . . . widely accepted among rate authorities. . . . But thirty years ago it was a bold departure from custom; its results were uncertain. . . .
>
> The basic idea behind the promotional rate structure is this: the greater the load density on an electric distribution system the greater the economy of operation and use of materials; the larger the demand for power the greater the opportunity of developing large power resources and the greater the economies which come from generating on a large scale. These factors lower the cost of power to consumers.
>
> The promotional rate has played an important part in enabling the Commission to break the circle of high price and low consumption. . . . The Commission tied power charges to costs and kept costs low. It cut rates at every opportunity, consistent with maintaining a sound financial position. *The demand for electricity was found to increase rapidly as its price was reduced.*

Without doubt the promotional rate structure contributed importantly to the phenomenal increase of Hydro's power load. In 1912 the municipal partners had 33,568 lighting customers and 1,399 commercial power users. In 1913 the figures were respectively 63,157 and 2,532—almost double. By the end of 1913 Toronto Hydro had 22,320 customers and a peak load of over 22,000-hp.

Not satisfied with this booming expansion, in the summer of 1912 Beck put his famous Hydro Circus on the road. Like the promotional rate structure it was a Hydro innovation which helped greatly in speeding the extension of electric service in rural districts of Ontario.

The Circus was in two units, each consisting of a caravan of two horse-drawn covered wagons, one carrying a motor and cables for connecting with power lines, the other a pair of 15-kilowatt transformers for stepping down power from 2,200 volts to 220 or 110 volts. The loads were so heavy that frequent changes of horses were necessary, as in stage coach days. Later a three-ton 50-hp truck was acquired. It carried the Circus personnel of speakers and demonstrators and a collection of electrical equipment for farms and farm kitchens, from a saw rig to a washing machine. Posters and flyers listed what the average electrified farm should have:

> Complete lighting system, using 25-watt and 40-watt lamps; a 500-watt flat iron; perhaps a vacuum cleaner and electric stove.
> Cow stable: row of lights behind cows, about one 20-candle-power lamp to every three cows.
> Horse stable: three lamps or four 20-cp lamps.
> Silo: one lamp.
> Drive shed: one lamp.
> Yard: one good 100-cp lamp on pole.
> Power: 5-hp motor, fixed permanently in barn to run line of shafting, or mounted on truck so that it may be moved from place to place—e.g., to a pump in the barnyard, where arrangements for connection have been made.

The private power companies had never indulged in promotional extravagance of this sort. Their concern was profit making and it didn't pay to serve isolated farms and communities in the sparsely populated areas. In the beginning it didn't pay Ontario Hydro and its municipal partners either, but it was necessary if Beck was to fulfill his promise to bring the life-giving elixir of cheap power to every Ontario hamlet and homestead.

Farmers, as Ontario Hydro's Chairman was acutely aware, were also voters. If they could be shown that Hydro offered them a real hope, and their only hope, of benefiting by the electrical revolution that was transforming the cities, then the rural legislators in Queen's Park might look kindly upon the idea of subsidizing the extension of rural transmission lines which the municipalities could scarcely afford to pay for.

But in 1912 they still had to be shown. More than a year before, in March 1911, legislation had been passed which permitted one or more ratepayers to apply for power without vote of other electors, thereby enabling individuals and the small unorganized communities known in Ontario as "police villages" to acquire municipal status so far as Hydro was concerned. But while farmers throughout the province had responded with interest, their numbers were insufficient to justify the adoption, or as yet even the formulation, of a comprehensive plan for rural electrification. As the wooing of municipal ratepayers had proved essential in the earlier phase of the public-ownership movement, so it had become necessary to woo the support of the farming elements if the boon of electricity was to become available to every section of the province.

At the time, however, rural electrification in both Canada and the United States was still pretty much a dream, so much so that Beck's European tour had been inspired by a desire to secure factual information. As one picture is worth a thousand words, so one demonstration, as Beck well knew, is worth at least a hundred pictures, and so the Circus had its origin.

Chugging their way along dusty roads, testing every bridge and culvert as they came to it, the speakers of the Hydro Circus, including often the magnetic ringmaster himself, moved from farm to farm and village to village, showing that the electrical dream of better living could be made a reality at cost. Whenever, as a result of their persuasion, the public will of some rural community like Beachville passed the Hydro by-laws, Ontario Hydro's Chairman would appear at the switching-on ceremonies. Bowler-hatted, genial, informed and endlessly informing, nobody could beat the drum for Hydro quite so well as the Chairman. Exhibits and demonstrations of household and farm appliances were standard features on such occasions. Food was cooked on an electric stove. A cow was milked by an electric machine and the cream separated in an electrically driven separator. Later in the summer, Ontario Hydro's electrical show was a major attraction at Fall Fairs at London, St. Mary's, New Hamburg, Stratford, Norwich, Woodstock, Weston, Tillsonburg, Markham, Dundas, and the Toronto Exhibition.

Beck's Circus had its first tryout on the Might Farm in Toronto township on August 28, 1912. Present at the speakers' table were Beck, J. S. Duff—the Ontario Minister of Agriculture—and township officials. The first caravan sent out was conducted by James W. Lotimer, who had set up the electrical exhibition, and J. W. Purcell, who knew farming and farmers. For a while H. D. G. Crerar left his testing laboratory to tour with the Circus in Oxford, Norwich and Durham townships.

The Circus was divided into two sections, one under Crerar which visited the county fairs, while the other, under J. W. Purcell, gave shows and demonstrations at the farms. They were self-contained units with their own transport, for the most part trucks, and tents. Crerar's section also had a small commissary service and the cooking was in charge of a home economist borrowed from the General Electric Company. Wherever it camped in a town, the civic dignitaries were usually invited to a party at which this lady gave demonstrations in the art of cooking by electricity, providing what were probably the first electrically cooked meals in Ontario.

That winter Crerar was back in his Toronto laboratory and in the spring of 1913 was off for Europe in company with F. A. Gaby, who had succeeded Sothman as Chief Engineer, to study electrical developments in Europe. Crerar's concern was with improvements in incandescent lamps and lighting generally; Gaby's with the advance of rural electrification.

The section under Lotimer experienced all the vicissitudes of barnstorming; the breakdowns on execrable roads; the frenzied efforts to keep going to reach the next stand on time; the catastrophes in ditches; the repairs made by weary men at night by the light of lanterns; the teams borrowed and the rousing of local blacksmiths from their sleep to improvise substitutes for broken parts. And not uncommonly at such times Beck would appear suddenly in the big Pierce Arrow which had become his favorite mode of transportation. On one such occasion, when things had reached their desperate worst he arrived, unrecognized, and began to shout directions. Whereupon he was told to go away and mind his own business—whatever that might be.

It was disclosed later that the Circus cost between $25,000 and $30,000, a heavy enough charge for those days but now a seemingly trifling amount in the light of the cost of television programs. In the fall of 1912 it was therefore abandoned in favor of demonstrations at local fairs and other less expensive promotional activities. Beck spoke at many of these demonstrations, and his eloquence improved with practice. In November of 1913, at a dinner given to honor the Chairman of Ontario Hydro and his wife by the citizens of London, Beck said:

> If I have helped to lessen the household cares of the housewife by making electricity her servant . . . if I have helped the farmer to make life on the farm more attractive, to help keep the boys and girls on the farm, then I have not labored, nor have you coöperated with me, in vain.

12

That Beck had not labored vainly, nor had the people of Ontario coöperated with him in vain, was becoming uncomfortably clear to the big corporate electrical interests south of the border in the United States—New York, Illinois and Oregon principally being involved.

Campaigns to discredit Hydro were launched by American public utilities and in 1912 a broad offensive was mounted, spearheaded by sympathetic members of the New York State Legislature. For many years the States had suffered from the excessively high rates of the private-power interests, but the national bias in favor of individual enterprise was to preclude the possibility of any important attempts at public ownership and operation of hydro-electric power resources until the advent of the depression-born Tennessee Valley Authority in 1932.

New York State was a good base for a preventive war designed to arrest the southward spread of the public-ownership virus. A legislative committee of eight, headed by Senator T. Harvey Ferris, was instructed to report on Ontario's public-ownership experiment. When the Committee arrived in Canada, it was accompanied by Reginald Pelham Bolton, a New York engineer and well-known private-utility partisan. Suspecting that the Ferris Committee meant Hydro no good, Beck resolved, apparently, that they should get nothing out of him or the Commission's staff. Pursuing this tactic, Beck repeatedly broke appointments with the Committee, and when the American delegation tried to get their questions answered by Ontario Hydro's engineers, they could get nowhere. It was obvious that Beck had told his staff to tell them nothing.

After a week of continuous rebuffs, the Committee left. When it returned in mid-June it was told by Beck and Gaby that there *might* be some information for them in November, after the end of the Ontario Hydro's year, but again there might not. During the summer Ferris pre-

108

dicted that the facts about Hydro, when published, would put an end to agitation for public ownership in New York State. Beck's continued refusal to meet with the Committee was therefore understandable. Its letters went unanswered and, in December, when the Committee returned, it learned from Gaby that Beck had left Toronto without authorizing him to furnish information of any kind about Hydro's operations.

In January 1913 the Ferris Committee issued its report. It found Hydro to be a colossal failure. Measured by ordinary business accounting, said the Committee, Ontario Hydro was running up a deficit of half a million dollars annually, and Toronto Hydro a deficit of a quarter million. The smaller communities were being grossly overcharged for the benefit of Toronto, Hamilton and London, where Hydro customers were saving $652,000 a year. Capital costs of $8,000,000 constituted a staggering burden on the municipalities and on the Province; and all this to develop a paltry 30,000-horsepower, when New York State needed 2,500,000-hp. As for Hydro's idea of developing the international section of the St. Lawrence, such talk was arrant nonsense.

So went the majority report. But there was also a minority report, signed by Committeeman J. L. Patrie. Patrie declared that the majority had based its report on data obtained through Bolton from a dismissed Ontario Hydro official, and that this data was contradicted by everyone else whom the Committee had interviewed in Canada, including even private-utility managers.

Was it conceivable, asked Patrie, that the development of hydro-electric power really was uneconomical, as the majority alleged? The General Electric Company, largely owned by such astute capitalists as J. P. Morgan, E. T. Stotesbury, Charles Steels, and George R. Sheldon, controlled nearly half the commercially developed water power in the United States. These gentlemen, surely, were not in business merely for their health.

Patrie accepted Hydro's accounting, which showed that it was saving the people of Ontario about $3,000,000 a year, and thought the postponement of payments to its sinking fund was entirely reasonable. Ontario Hydro's rates to the municipalities were about one-third those being charged New York State consumers by private companies. If this was what happened under public ownership, concluded Patrie, New York would be well advised to follow Ontario's example.

The New York State Conservation Commission, which issued a report on Ontario Hydro about the same time, agreed with Patrie. With only a single dissenting voice, the Commission reported that Hydro was soundly financed and efficiently operated; that its service was reliable and its rates low and getting lower. It recommended also that the State of New York follow Ontario's lead. This recommendation was also supported, in vain, by Governor Sulzer and Lieutenant-Governor Glynn.

Soon after the Ferris Committee reported, its chief informant, Reginald Pelham Bolton, published a book of 280 pages, entitled *An Expensive Experiment: The Hydro-Electric Commission of Ontario*. The book bore all the earmarks of propaganda, bought and paid for by the private-power interests, although its author declared that no one had paid or promised him anything. It was the forerunner of numerous publications of the same kind which appeared during the next two decades directed against publicly owned utilities and in particular against Hydro. Concerning his sources, Bolton wrote in his book:

> The citizens and taxpayers of the State of New York, confronted with a proposal for the expenditure of their money in an evidently ill-considered and distinctly ill-informed scheme of government speculation in the electrical industry . . . turn to a sister Province for information as to the working of a governmental experiment of the character proposed for their adoption. . . . They were met by disregard and neglect, when a fortunate circumstance, due to the disgust of one honest man with the methods and secrecy of those in command of the affairs of the experiment, brought to their service the actual records upon which the disclosures herein were made possible. The aid rendered by that Honest Man without regard or remuneration can only be acknowledged by this statement since he has determined to refuse publicity for his service to the public.

Bolton did not name his "Honest Man," but the publicity for the book featured a photograph of P. W. Sothman, Ontario Hydro's one-time Chief Engineer, who, rather than put his honesty to trial, had fled the jurisdiction of the Ontario Courts.

The press of the border city of Buffalo participated energetically in these across-the-border attempts to strangle Hydro in its cradle. The motivations of the anti-Hydro newspapers were exposed when Buffalo's City Council authorized its Corporation Counsel to seek relief from the extortionate rates Buffalo was being forced to pay its privately owned utilities: $25 to $38 per horsepower for the same Niagara power that Toronto was getting from Ontario Hydro for $15 per hp. The relative transmission distances were twenty and eighty miles.

With one exception, the *Buffalo Express*, the city's newspapers refused to print anything against the private power companies. Instead, they editorialized vociferously on their behalf and put out pamphlets charging that Ontario Hydro was losing $3,000,000 a year. Three of these newspapers, the Corporation Counsel discovered, had for years paid nothing to the private power company for electrical service. When these facts became known a court order forced the company to collect the unpaid accounts.

While these charges were being made, Beck, on his way to England, interrupted his journey in Albany to tell Governor Sulzer and Lieutenant-Governor Glynn the actual facts about Hydro. He told them how rates had been cut in all the Hydro municipalities; how all the municipalities had met their obligations to the Province without incurring any deficits; and how the ratepayers had first approved Hydro by impressive majorities and lately by almost unanimous votes. He spoke also of line construction works that had been completed far below estimated cost, and of the rapid increase in the number of communities serviced, including many small villages which, without Hydro's low costs, would still be without electricity. He denounced Bolton's book as mendaciously conceived, grossly erroneous, and in part sheer fabrication.

Meanwhile the Canadian private power companies continued their efforts to undermine, obstruct, and generally contain the public-ownership movement. Communities were discouraged from joining Hydro, by the offer of low-cost, long-term contracts and other inducements. In North Bay, the city which had become the gateway to Northern Ontario with its great mineral and forest resources, a heated campaign was waged over the public versus private power issue. In December 1912 the ratepayers voted on two questions: one, the approval of a standard Hydro enabling by-law; the other, the renewal of the franchise held by the North Bay Light, Heat and Power Company, the subsidiary of Electric Power. The vote for Hydro was 561 with 88 against; for renewal, 142 for and 679 against.

In the second round of this same battle, the victory was less complete. Called upon to approve a money-law for $60,000 for the power plant of the private company, about one hundred ratepayers changed sides to make the vote 598 in favor, 213 opposed. Anti-Hydro partisans on the North Bay Council, claiming that "a change of opinion was apparent," refused to sign a Hydro contract. The situation prevailed until 1916 when the Electric Power Company and its numerous subsidiaries were purchased by the Provincial Government and control and operation were vested in Ontario Hydro.

Beck might possibly have been able to win this and other skirmishes on Hydro's eastern frontier, had he not found it necessary at this time to put down a rebellion against his authority led by P. W. Ellis, Chairman of Toronto Hydro. The unity of the public-ownership forces was further impaired when a mutiny against Ellis exploded in the ranks of Toronto Hydro itself. As the head of Toronto Hydro, Ellis had to establish his authority over his own staff, and as head of Ontario Hydro, Beck had to establish his authority over Ellis and Toronto Hydro. The resulting controversy precipitated another of the bitter and exciting political battles in which Ontario Hydro's Chairman was constantly involved.

The insurrection against Ellis came first. In 1912, after K. L. Aitkin was forced to resign as General Manager of Toronto Hydro because of ill health, Ellis and his co-director, Mayor Hocken, appointed W. R. Sweany as Acting-General Manager. He had been with Toronto Hydro only three years and Ellis believed a more experienced and better-qualified executive was needed for the post. After lengthy inquiries at home and abroad, the choice rested on H. H. Couzens, the manager of the electrical utility in the Borough of Hampstead, London, England. Early in 1913, a three-year contract was signed with Couzens to manage Toronto Hydro at $10,000 a year, an unusually high salary for the period. The appointment was not to become effective for some months and was not mentioned at any regular meeting of the Commissioners. However, news of it leaked out.

Angered by being passed over, Sweany set out to undermine Ellis, whose appointment by the Toronto City Council extended only to June 1. In this effort, the engineer was supported by the politically powerful Toronto *Telegram* and its able chief editor, John R. Robinson. Although the paper had always been an ardent supporter of Beck and public ownership, it cordially disliked Ellis for his indifference to the paper's proposals, and as the standard-bearer of the Ontario Ulster and Orange community exerted a great influence on civic and provincial public opinion. Joining forces with Sweany, the paper endorsed the submission of a secret petition, signed by him and ten top executives of Toronto Hydro, to the City Council, urging the non-renewal of Ellis' appointment on the grounds that he was unduly subject to political influence and otherwise unsuited for his responsible position.

When Ellis heard of this he promptly fired Sweany and gave his co-conspirators their choice of renouncing Sweany or being themselves dismissed. When they refused, Ellis made good his threat, but reinstated half of them when they recanted. He then announced the appointment of Couzens as General Manager, following which his own reappointment was voted by the Toronto Council.

But Ellis' enemies still did not admit defeat. Inspired by the *Telegram*, two members of the Board of Control introduced a proposal to have Toronto Hydro investigated by the Board. The attempt failed when Beck refused to approve such an investigation and when Ellis made it clear that he would resign if an investigation was ordered. In a public statement, Ellis declared:

> If Commission government is to be subject to investigation upon the demand of intriguing employees, it is at an end. No honourable and capable man will devote his time to the onerous duties of such work if he is to be harassed by investigating commissions appointed under such circumstances. I desire to place on record the principles of administration for which I stand so that there may be no mistake on anybody's part regarding them. They are very simple, namely:
>
> 1. That all appointments and promotions shall be made to secure the highest practical efficiency in every department, without regard to the religion, politics, or personal influence of any man.
> 2. That the safeguard of an absolutely independent and unrestricted audit of income and expenditure to protect the public moneys shall be maintained.
> 3. That truthfulness of accounts and integrity toward the public in all relations shall be constantly observed.
>
> I am not a politician nor a lobbyist. I am a business man. And I will not serve Hydro or any other system except upon the basis indicated.

Pending Couzens' arrival from London, Ellis and Mayor Hocken appointed as Assistant General Manager R. G. Black, a former employee of the Mackenzie Syndicate who, eight years before, as President of the Canadian Electrical Association, had signed a critical appraisal of Ontario Hydro's estimates. Instead of being enraged by this appointment, as everyone expected, Beck enraged the Toronto *Telegram* by appointing Black to the long-vacant post of Provincial representative on the Toronto Hydro.

The appointment, as it turned out, did nothing to weaken the growing disposition of Toronto Hydro to manage its own affairs, with or without the benediction of Ontario Hydro's dictatorial Chairman. One of General Manager Couzens' first acts was to recommend the construction of a stand-by steam plant as insurance against service interruptions. Over a period of eighteen months there had been four 6-hour breakdowns, during which Toronto Hydro, to keep the City's water supply in operation, had been obliged to turn for help to the reserve steam plant of the privately owned Toronto Electric Light

Company. But construction of a reserve steam plant for the city would require the use of surplus funds which could otherwise be applied to a further reduction of Toronto Hydro's rates, which were already 50 per cent lower than those of the private company.

At this point, the smoldering conflict between Beck and Toronto Hydro flared into violent controversy. Beck wanted all surplus revenues to be applied to rate reductions—his most effective weapon in his fight-to-the-death with the private companies. To the remonstrances of Toronto Hydro, who wanted not only to protect their customers against service interruptions but also to build up a reserve against future bad years, Beck replied that there wouldn't be any future bad years. As for the need of a reserve steam plant, that was something for Ontario Hydro to worry about.

At an earlier date, and without consulting Toronto Hydro, Beck had announced, shortly after his visit to Albany to rebut the attack of the Ferris Commission, that he had recommended a 10 per cent cut in Toronto Hydro's rates. The Toronto *Telegram* had published this announcement, which Ellis and Hocken had read with surprise and unconcealed resentment. They were not prepared to reduce rates, said so, and were supported in their refusal by the Toronto City Council. In this attitude they remained adamant.

Beck's attempt to dictate the policies of Toronto Hydro lacked the support of his fellow-Commissioners, Hendrie and McNaught; the latter went so far as to say publicly that he regretted the manner in which Beck had announced a reduction of Toronto Hydro's rates. This led to talk of dissension within Ontario Hydro, which was aired by the Toronto *Telegram*. The paper had a dependable sympathizer in Controller Church of the Toronto City Council. Church introduced in the Council a motion that Toronto Hydro be investigated. When the motion was defeated by a vote of 18 to 2, Church then urged that McNaught and Hendrie be replaced in Ontario Hydro by new appointees, to be recommended by the Ontario Power Union, who would give Beck unquestioning support.

President Lyon of the Ontario Power Union thought this was a good idea and Commissioner McNaught was willing, but Commissioner Hendrie, never a Beck supporter, considered the suggestion outrageous and said so.

In the Legislature, the Liberal leader N. W. Rowell saw an opportunity to make political capital out of the dissension within Ontario Hydro and the conflicts between Beck and other Cabinet ministers in the Conservative Government. Somebody would have to go, declared the Liberal leader. These bickerings in high places must end, else the interests of the humble consumers who paid for Hydro and bought its power would suffer. This political oratory was not without its effect on Premier Whitney, who agreed with his Minister of Power on at least one point: the importance of Beck's support in the coming election. If Beck left the Commission and the Government, as he had threatened to do once before, the Conservatives would be obliged to contest the election without the support of Beck's huge personal following and his personal pressure group, OMEA.

The only alternative was to give Beck what he asked for: absolute authority to run the Hydro enterprise as he saw fit. McNaught and Hendrie, the Government's two nominees on the Commission, must be induced to support Beck's order to Toronto Hydro to comply with his recommendation to reduce rates.

In May, a month before the election, Beck had his way. A formal legal decree was issued "ordering and directing" Toronto Hydro to put into effect the 10 per cent rate reduction.

Whitney was ill, but his party was afraid to go to the polls without him or without Beck. Whitney called Beck to his bedside. Shortly before the election, said Whitney, the King's birthday honors list would be published. Beck's name was on the list. Would he accept this time the knighthood he had pushed aside once before when he had declined to have his name proposed? And if he accepted, would the Ontario Hydro Chairman display a little more interest in the election than he had thus far manifested?

What actually transpired in the pledges given and received remains unknown, but Adam Beck was awarded the title of Knight Bachelor in His Majesty's honors list for 1914, and on the eve of the election, at a great rally in Toronto's Massey Hall, Sir Adam Beck appeared beside the ailing Premier. With Whitney, the idol of the public-ownership partisans joined in an appeal to the voters to affirm their confidence in the Conservative Government on election day. The reunion proved effective. On June 29, the Conservative majority was 84 to 27.

Within three months Whitney was dead, and Beck was still faced by a defiant Toronto Hydro. The Whitney Government had done its best, but it had not been able to make Toronto Hydro change its position. Ellis and Hocken refused to comply with the formal order to reduce rates, although every day of non-compliance was costing a $100 fine.

For the City Council, Mayor Hocken marshalled his arguments against rate reduction in a 50-page pamphlet. The order of the Commission, he insisted, was a gross and unwarranted interference with the rights of Toronto Hydro by a provincial authority which had only limited financial responsibility for what might be done.

Cannily, Beck avoided invoking legal compulsion. Instead, he renewed, through the City Council, his pressure on Toronto Hydro. Again Controller Church demanded a judicial investigation of Toronto Hydro and again he was refused. A resolution that Toronto Hydro be requested to obey the Ontario Hydro order was defeated by a vote of 13 to 6. Toronto Hydro proposed a test suit to determine the respective jurisdiction of the two Commissions, but nothing came of it. Beck then suggested that the two Commissions arrange a joint conference. This met with immediate approval by Mayor Hocken. Shortly Ontario Hydro's Chief Engineer Gaby was conferring amicably with Toronto Hydro's General Manager, H. H. Couzens, but nothing was decided. A meeting between the two commissions in the Parliament buildings, from which the press was excluded at the insistence of McNaught and Hendrie, was equally without result. Beck denounced the exclusion of the press, scolded the Toronto Commissioners, and when told gently by Ellis to "stop splashing around," splashed all the harder.

Eventually the stalemate was ended by the outbreak of World War I and the growing power requirements of Ontario's war industries. In December of 1914 Toronto Hydro voluntarily reduced rates, not by the flat 10 per cent which Ontario Hydro had ordered, but by an average of 18.1 per cent to all categories of consumers. Toronto had a new mayor, none other than the *Telegram*'s henchman, Thomas L. Church, who by his election to that office became *ex-officio* a member of Toronto Hydro. With notable assistance from Church himself and the reinforced support of the *Telegram*, Beck had again outmanoeuvred his opponents.

13

Under the forced draft of Canada's industrial mobilization during World War I, Hydro entered upon a new phase of feverish expansion. Whitney was dead, and at last Beck had a Government and Commission made more or less to his order. The new Premier was William H. Hearst of Sault Ste. Marie. As Minister of Lands, Forests and Mines he had occupied a key post in the Whitney cabinet during a period of unprecedented development in Northern Ontario. Begun with the building of the government-owned Temiskaming & Northern Ontario Railroad by the Ross Government and accelerated by the discovery of silver at Cobalt, industrialization had entered a new phase, when advancing technology made possible the use of the great hydro-electric power potential of the North for the exploitation of Ontario's vast but hitherto useless supplies of pulpwood. Cautious and conservative in nature, an administrator rather than a politician, Hearst had never opposed Ontario Hydro's Chairman in the past, and presumably would collaborate in the wartime fulfillment of Beck's soaring ambitions. Beck, by his own wish, was not named a member of the cabinet. But neither was his old enemy, John M. Hendrie. Knighted at the same time as Beck and now Sir John, Hendrie had been elevated to the post of Lieutenant-Governor, a position which, in theory at least, removed him from the storms of politics. Remaining was W. K. McNaught, a staunch Beck supporter despite his defection during the Toronto Hydro rate-reduction controversy. To the vacant chair at Ontario Hydro's Board Room table Premier Hearst appointed the new Provincial Treasurer, I. B. Lucas, a loyal supporter of both Beck and Hydro.

With W. W. Pope continuing as Secretary, Fred Gaby as Chief Engineer and Harry G. Acres in charge of the Hydraulic Engineering office, Ontario Hydro had a strong wartime administration team, with plenty of talent coming up through the lower echelons. Thomas H. Hogg, the

117

Assistant Hydraulic Engineer, was later to become one of a succession of distinguished Ontario Hydro chairmen, as was Richard L. Hearn, the design draftsman. Two other brilliant young engineers, Otto Holden and J. J. Traill, found in Hydro's wartime expansion a chance to prove their professional capacity. Known as "Beck's bright young men," they worked interminable hours, met and solved novel engineering problems, and more than earned the accolades that were later bestowed on them in world engineering circles.

Ontario Hydro's engineering team had no time to waste on administrative protocol and no money to spend for elaborate equipment. When Beck dropped into the Hydraulic Division office to see how things were going, he would find the designs and working drawings for major undertakings laid out on plain kitchen work-tables: generating stations, rural electrification extensions, and power systems to serve the mining, lumbering, and pulpwood centres of the forbidding hinterland that reached north and westward to Hudson Bay and the boundaries of Manitoba. Also on the wartime drafting boards were preliminary plans for two great schemes still in the discussion stage: Hydro radials, and a canal from Chippawa to Queenston that would use Canada's full share of the hydraulic resources of the Niagara River to supply the world's largest electrical development.

Whether epic or routine, the work of planning was only a part of the day-by-day work of the Commission. Included also were complex problems of engineering and construction, the continual expansion of the municipal partnership, the negotiation of contracts, the operation and maintenance of a rapidly expanding plant, and the procurement of supplies and materials in a wartime market. There was also the ever-present need to keep the political fences mended in the interests of public-ownership support. During the period Ontario Hydro's limited reserves of trained technicians were constantly depleted by the departure of valuable staff members for war service.

In an emotional climate which banned Beethoven, ordered Salisbury steak instead of Hamburger, and changed the name of Berlin to Kitchener, it was perhaps inevitable that among Beck's enemies were some who tried to injure him because of his German ancestry, although this was several times removed. As a matter of historical record, Beck cabled the British War Office the morning after war was declared, offering his services in any capacity to which he might be assigned. Two years before the war, as one of Canada's leading horsemen, he had done remount work for the Army and was now made an Honorary Colonel in charge of cavalry remount. To help meet the early and critical shortage of horses, Beck gave the British Army ten from his own stables. One of them, a prize-winning animal named Sir James, was ridden by General Alderson, head of the Canadian divisions on the Western Front.

As director of remounts, and a member of the Canadian Army Head-quarters Staff, he also supervised the work of fifteen horse-buyers in the area between Lake Superior and the Atlantic and later set up horse depots in every part of the Dominion.

Shortly after the war started, Canadian manufacturers were asked by the Dominion Government to what extent they could assist in sup-plying the urgently needed munitions. While many were willing to con-vert their entire facilities to munitions production, they could not have operated at full capacity if the Commission had not been able, during the first three years of the war, to supply these plants with large quan-tities of power.

In 1914 the Commission for the first time in its history became a producer of power, in addition to transmitting it. In July the 5,600-hp Big Chute GS also on the Severn River, was purchased from a private company. It served Collingwood, Barrie, Midland, Penetanguishene, and the surrounding districts. The 1,200-hp development built by Ontario Hydro at Wasdell Falls on the Severn River was placed in operation in October to serve the Beaverton-Canning district. In addi-tion, construction began in 1914 at a power site purchased by the Com-mission at Eugenia Falls on the Beaver River. The objective was to de-velop 8,500-hp for Owen Sound and its district. The small systems served by these three pioneer stations were later to become intercon-nected with a number of others to form the Georgian Bay System and, later still, the Georgian Bay Division of the Southern Ontario System, but during the 1914-18 period they were entirely separate from the Niagara System, where the greatest concentration of the munitions in-dustry was located.

In the latter system, the Commission had contracted with the Ontario Power Company for a supply of power up to 100,000-hp. By October of 1914, after just four years of operation, the system load had grown to 71,994-hp. A reserve was therefore still available for the load growth that was likely to arise in plants engaged in war work. Ontario Hydro was able to satisfy the requirements of its industrial customers during the first year of the war. By November 1915, however, with industrial production being steadily stepped up, the reserve was exhausted and it became necessary to purchase power from the Toronto Power Com-pany and Canadian Niagara Power Company plants at Niagara Falls. Purchases from these sources, along with the virtually continuous operation of the Ontario Power Company plant at overload, enabled the Commission to take care of war-inflated loads on the Niagara System, which were to grow by October of 1918 to 205,912-hp, almost three times as large as in October 1914.

In August of 1917, the generating plant, transformer stations, transmission lines, and sub-stations of the Ontario Power Company were purchased by the Commission with a view to improved service and increased economy. Plans were made to increase the capacity of the generating station by an additional 45,000-hp. Under great difficulties because of shortages of labor and materials, the extension was rushed to completion early in 1919.

At the time of the signing of the Armistice, the Niagara System was supplying over 80,000-hp to 360 plants working on the manufacture of munitions. Within eight months other industries had absorbed all the power made available by the cessation of war. So rapid was postwar industrial growth that before the end of 1919, it was necessary for the Commission to limit the amount of power supplied to its municipal customers.

Not all of the munition plants were in the districts served by the Niagara System. In 1916 the Provincial Government had purchased the system of the Electric Power Company in Central Ontario, placing it under the control of the Commission, which then set about to improve services to munitions plants in the Oshawa and Peterborough areas and in other districts east of Toronto.

As the military, agricultural, and industrial demands of Canada increased, notwithstanding the total manpower available, skilled labor had moved into short supply and along with it all kinds of replacement parts, while those usually obtained from foreign sources had by now become unprocurable. In consequence, recurrent plagues of line breakdowns tested the ingenuity of an overworked staff and the patience of Hydro's customers.

One such failure had occurred a year before the war, during a hot spell in June 1913. It lasted six hours and forced Hydro to turn for help to its rival, the Toronto Power Company, in order to keep the city's water system in operation. Another major breakdown, caused by heavy winds and icing rain, occurred in February 1914, when over a hundred passengers on the London and Port Stanley Railway were stranded for hours in freezing coaches. In the Scott Opera House in Galt, a play was performed by candlelight; in Toronto during a production of *Samson and Delilah* an audience at the Royal Alexandra Theatre waited in darkness until a connection with the Toronto Electric Light Company enabled the show to go on.

When these and similar emergencies arose, all the engineers at head office were assembled and briefed on the nature of the disruption. Their duties were to supervise the work of the line crews on the various sections of the high-voltage system where trouble had been reported, travelling by train to the stations closest to the affected area, then proceeding by horse and wagon. There were then no Ontario Hydro trucks. On the job a complete inspection was made of all the suspected insulators (of which there were then 125,000 in use) on the 115 kv lines, the testing being carried out by groups. Where insulator units had to be replaced, it was necessary to turn off the power. The cycle of reactions that might be started in suspension insulators by abnormally high temperatures was at first not clearly understood. The different expansion characteristics of the porcelain and the metal in the insulating units, as well as many other items contributing to flashover and power interruptions, required considerable study and it was many years before all the problems involved were finally solved. Fully satisfactory insulators were not manufactured until about 1918.

Again and again the heat lightning of public protest, generated by a succession of service breakdowns, played about the exposed head of Ontario Hydro's Chairman. Although he could easily have passed the blame along to his engineers, he defended them warmly on every occasion. They were not to be blamed, he insisted, either for wartime shortages of men and materials or for the occasional failure of apparatus that was still being perfected. Ontario Hydro was doing its best, and its best was steadily getting better.

This was patently true. Ontario Hydro's rapid expansion both before and during the war was accompanied and facilitated by periodic technical advances that helped to improve both service and public relations. Between 1910 and 1920, Ontario Hydro extended its transmission lines from 523 to 3,332 circuit miles; the number of transformer stations increased proportionately. From a meagre 4,000 kw in 1910, when Canada entered the war in 1914 the demand for Hydro power increased to 68,000 kw, only 5 per cent of which was provided by the Commission's own plants. By December 1918 the demand was 221,300 kw, and of this tripled power load the Commission's own plants were supplying 65 per cent.

The Commission's second development, at Eugenia Falls on the Beaver River near Georgian Bay, was started in July 1914. The first section was completed in November 1915, the remainder in 1920.

In 1916 the Ontario Government, as mentioned, had purchased the assets of the Electric Power Company, the merged corporation in Eastern Ontario whose ramifications included a small generating station serving the northern communities of Nipissing, Powassan, Callander, and North Bay. Ontario Hydro operated the Nipissing System for the Province until 1928 when it became, for purposes of financial administration, part of the Commission's Eastern Ontario System for four years, after which it formed part of the newly organized Northern Ontario Properties.

As early as 1901 the lakehead city of Port Arthur had constructed on the nearby Current River one of Ontario's first municipally owned generating plants. Later, after Port Arthur had voted for public ownership, Ontario Hydro bought power from the local Kaministiquia Power Company to supply increased demand. Later still, when this service was no longer adequate, Port Arthur and its lakehead twin, the City of Fort William, entered into agreements with the Commission for a larger supply of power. Surveys disclosed that the only source in quantities sufficiently large to satisfy the demands of the burgeoning mining and pulp-and-paper industries of the Thunder Bay district was the Nipigon River, some 125 miles away. Known to sportsmen the world over for its fabulous speckled trout, the river fell 248 feet in the 32 miles between Lake Nipigon and Lake Superior and was therefore exceptionally adapted to power development.

Cameron Falls, with a drop of 72 feet and a storage area of 12 miles extending north to Portage Lake, was chosen for the first development. Situated in a trackless, virtually uninhabited wilderness, accessible only by the main transcontinental line of the Canadian Pacific Railway, the site presented unique problems in engineering and construction. To solve these, roads had to be built across rugged Pre-Cambrian formations; construction camps provided at two dam sites, Cameron Falls and Alexander Landing, two miles down river where the fall is another 60 feet; and deep-water facilities constructed at Lake Superior. Begun in 1918 and completed in 1920, when power was delivered by two 110,000-volt lines to Port Arthur and Fort William, the first two Northern Ontario plants were the experimental laboratories which enabled Ontario Hydro's engineers to pioneer notable advances in hydro-electric technology, particularly those which were to be applied later to the development of Canada's even more remote hydraulic resources.

Another notable technical advance, though dissociated from the peculiar difficulties of the northern region, had to do with automation, one of the most revolutionary steps in modern industrial production. On taking over the plants of the Electric Power Company in 1916, Ontario Hydro found itself in possession of a number of relatively primitive installations on the Trent Valley Canal System, a much-derided pork-barrel undertaking to connect Georgian Bay with Lake Ontario by way of Lake Simcoe and the Trent and Severn Rivers. The Federal

Department of Railways and Canals had constructed storage dams and locks, and the Electric Power Company had installed small generating stations. To increase the efficiency and the economic operation of these and additional units built there by the Commission, Ontario Hydro engineers invented and installed controls and switching gear, actuated by electrical impulses, which enabled generating plants to respond automatically at the direction of a station operator several miles away. Completed in the mid-1920's, these installations pioneered techniques which later made possible the operation of remote generating plants without the use of human hands.

In 1917 the Commission bought out the Ontario Power Company's generating station at Niagara Falls, with a present-day capacity of 135,000 kw, at a cost of $18,485,000, and at the same time embarked on the construction of the great Queenston-Chippawa GS (the name of which was changed in 1950 to Sir Adam Beck-Niagara Generating Station No. 1), plans for which were begun in 1914. It also began negotiations for the 108,000 kw station of the Toronto Power Company, in operation since 1906, but the purchase was not finally consummated until 1922, after prolonged negotiations with the Mackenzie syndicate.

Meanwhile the war had ended, and Canadians were returning home —many former Ontario Hydro employees who had been promised that their jobs would be open for them on their return. Most of Ontario Hydro's soldier employees returned to the fold and by the late spring or early summer of 1918, the reinforced organization was again in a position to undertake the stupendous tasks that lay ahead of it. Many, however, did not return. Their names are commemorated in stone at the entrance to Ontario's Hydro's Head Office on University Avenue in Toronto.

THE LARGEST HYDRO-ELECTRIC
PLANT IN THE WORLD

14

In the records for Ontario Hydro for 1914 appears the figure
of $6,354 itemized "Niagara Surveys." From this it is clear,
even if other evidence were lacking, that the great Queenston-
Chippawa GS had already become something more than a gleam in the
eye of Sir Adam Beck.

In point of fact, a greatly challenging and exciting engineering project
was afoot: the planning of a hydro-electric development which for the
first time would fully exploit the fall in the Niagara River from above
the Falls to its discharge from the Gorge between Queenston and Lewiston, instead of, as until then, using only the drop occurring at the Falls
themselves. This involved the design and construction of a generating
station to eclipse any previously built at Niagara and larger than any
yet contemplated anywhere in the world. As tentatively envisaged, this
new plant, to be erected at Queenston three and one-half miles down
river from the Falls, would in a single daring leap advance the technology of electrical production far beyond anything ever attempted.
Unprecedented engineering problems would be met, and the cost would
be staggering in contemporary terms. But there was no doubt in the
minds of those close to Hydro as to the eventual completion of the
project. Sir Adam Beck was already committed to it.

By 1915 the twelve original municipalities to which the Commission had supplied electrical service in 1910 had increased in number
to 104. From an initial 1,000-hp, Ontario Hydro's power load had
risen to well over 100,000-hp, the limit of the amount stipulated in the
contract with the Ontario Power Company, and the Commission had
been obliged to make a temporary contract with the Toronto Power
Company for an additional 16,000-hp.

It was the inescapable implications of the ascending power curve
that first prompted Beck to start on the preliminary surveys for a new
Niagara development. As yet, however, the publicly owned enterprise

124

was limited by its contracts with the coöperating municipalities to the transmission of power bought from privately owned producers. As early as 1903 the original Snider-Detweiler report had foreseen the future need to manufacture power as well as engage in its transmission and distribution, but that contingency had been deemed so remote as to call for no more than a passing mention. To proceed with the gigantic Queenston-Chippawa project, therefore, it was necessary to obtain additional legislation, in support of which Beck again had to constitute himself a one-man lobby before a Cabinet of which he was no longer a member.

Even prior to the provincial elections of 1914, the Ontario Hydro Chairman had reportedly elicited from the ailing Premier Whitney the assurance of his support for the Queenston-Chippawa project. A year later, on November 15, with the mounting demands of wartime munitions plants to back him, Beck brought pressure on the Hearst Government to give the same assurance. Supported by Chief Engineer Gaby, he told the Cabinet that the Queenston-Chippawa development would take three years to build, by which time, if the war continued, the Commission would be running out of power. Since no other site was then available, it was essential that the work be authorized immediately. Beck estimated the cost at $20,000,000, this sum to include the construction of diversion works above the Falls, the building of a canal from the Welland River to Queenston, and the erection of a power plant to generate an initial 100,000-hp. By October 31, 1915, the total money advanced by the Province to the Commission amounted to just over $12,300,000. For his part, Premier Hearst, without disputing Beck's contentions, insisted that any project involving so large a sum must be put to a vote of the people. Confident of ultimate approval, Beck continued his preliminary surveys and, in 1916, to help prepare the manufacturers of heavy electrical equipment for eventual bidding, he issued preliminary specifications for the generators and transformers that would be required for the Queenston-Chippawa GS.

Despite the urgent need to meet Canada's mounting war commitments, a whole year elapsed before the ratepayers of the coöperating municipalities had an opportunity to cast their votes for or against the proposition: would they or would they not authorize Ontario Hydro to become a primary producer? When presented, the enabling by-law was couched in general rather than specific terms: it simply authorized the municipalities "to develop or acquire, through Hydro, whatever works may be required for the supply of electrical energy or power." No mention was made of Queenston-Chippawa, but no newspaper reader in Ontario could have remained ignorant of the purpose of the plebiscite to be held on New Year's Day, 1917. During the preceding months the province had once again become embroiled in an impassioned ideological campaign, for and against the Queenston-Chippawa development, into which Beck threw all his resources of organization, eloquence and manpower, while his opponents did everything they could to defeat him. Every town and city council, every municipal Hydro commission and the public at large had to be contacted and convinced of the advantages the enabling by-law would bring, for which purpose meetings were held nearly every night and addressed by Sir Adam or his engineers.

When the votes were counted after the New Year's Day plebiscite the results showed that in all municipalities except Goderich and Waterford the by-law had won overwhelming approval, which Premier Hearst accepted as a sufficient mandate to proceed. The next day he promised to introduce a bill in the Legislature defining the project as a municipal coöperative enterprise and empowering Ontario Hydro to issue bonds, to be guaranteed by the Province, to cover the cost of construction. The Act was duly passed and in May 1917 construction was begun on the most formidable power project ever undertaken up to that time.

An earlier scheme, promoted by American and Canadian commercial interests, for which the Ontario Power Company obtained a charter from the Dominion Government in 1887, had proposed "a canal from the Welland River or Chippawa Creek to the Niagara Gorge." Beck's engineers had taken over this idea and extended it imaginatively with the object of maximizing the theoretical potential. Previous power developments at Niagara Falls had utilized only that portion of the total fall which occurs in the vicinity of the Falls themselves, and even that relatively small portion had not been utilized to the full. In contrast, Ontario Hydro's engineers planned to utilize the greatest possible amount of the total fall of the Niagara River between Lake Erie and Lake Ontario. Of this total fall of 326 feet, about 12 feet occurs in the Upper Niagara River between Lake Erie and the mouth of the Welland River at Chippawa, and in the lower river from Queenston to Lake Ontario. It followed, therefore, that the maximum fall obtainable—314 feet—would be at Queenston at the crest of the Niagara escarpment. To convey the water to that point, however, would require a canal with a drop of 20 feet.

Subtracting 20 feet from 314 leaves a fall of 294 feet available at the powerhouse under maximum load conditions. This meant that 29.6-hp would be developed for every cubic foot of water per second that flowed through the canal. A measure of the enormous technical advance pioneered by the Queenston-Chippawa project is provided by comparing it with the three earlier plants built on the Canadian side of the Falls and the two on the American side. The most efficient of the plants on the Canadian side obtained only 17.1-hp for each cubic foot of water per second; each of the plants on the American side obtained only 9.5-hp. Thus every cubic foot of water to be used at the Queenston-Chippawa GS would produce between two and three times as much power as when it was used in the immediate vicinity of Niagara Falls.

To realize the greater power potential, the water of the Niagara River had to be diverted above the Falls and Rapids by a special intake structure built in the Niagara River at the mouth of the Welland River. The flow of the Welland would be reversed and the water thus diverted would pass along its deepened and enlarged channel for a distance of four miles. It would then enter the canal proper and traverse the Niagara peninsula for a distance of eight and three-quarter miles, passing through an earth section, then into the rock-cut section of the canal through a control gate of 48 feet clear span. The canal would be 48 feet wide and lined with concrete, the depth of the water from 35 to 40 feet. At one point near Lundy's Lane the floor of the canal would have to be more than 140 feet below the surface of the ground. The canal would terminate in a forebay to be situated at the edge of the Niagara Gorge, near Queenston.

On the edge of the Gorge, 320 feet above the river, a screenhouse would be built through which water from the canal would pass into steel penstocks, 14 to 16 feet in diameter, and be conducted down the face of the cliff to the turbines in the powerhouse.

The powerhouse itself, which by a series of revisions in the general plan was to have a generating capacity of over 500,000-hp, was to be built at river level against the face of the Gorge: an immense structure 590 feet long and 180 feet in height, approximately that of an 18-storey building, or higher than any commercial structure then existing in the British Empire. Planned on the scale of the great European cathedrals and not unlike them in the impressive beauty of its functional design, what was to prove a temple of energy would house nine generating units with capacities ranging from 55,000 to 65,000-hp, each under a head of 294 feet, and operating at a speed of 187.5 revolutions per minute. The total capacity of the mammoth development was estimated at 550,000-hp, and the time for the placing in service of the first units at three years. Actually, it took four years and a half from the start of excavation to the official turning-on of the first unit.

Construction of the Queenston-Chippawa Canal and Powerhouse

was started in May 1917 and the generating station was officially opened in December 1921. In spite of more than two years of the most careful hydraulic studies and engineering surveys, it was soon apparent that the project would take longer to build and would cost far more than had been originally estimated for the 100,000-hp initial installation. The studies included the building, under the direction of Professor R. W. Angus of the University of Toronto, of a scale model of the diversion works and canal for the purpose of obtaining hydraulic data unobtainable on the ground or from theoretical computations.

General policy and procedure for the power project was directed throughout by the Chief Engineer, F. A. Gaby, while the Chief Hydraulic Engineer, H. G. Acres, had complete charge of design and construction, with the exception of the bridges, which were designed by the Railway Department. For the greater part of the time Acres was assisted directly by R. L. Hearn in his capacity as Personal Assistant to the Chief Hydraulic Engineer. Thomas H. Hogg, the Assistant Hydraulic Engineer, gave detailed direction to the work of the various sub-divisions of the Hydraulic Department, including the work of the Hydraulic Engineer of Design, M. V. Sauer, and the Mechanical Engineer, Adolph Aeberli.

Henry Girdlestone Acres, a diploma graduate in Engineering from the University of Toronto, 1903, joined Ontario Hydro in 1906 as engineer in charge of the general water-power survey of the province. Previously he had held the position of Assistant Mechanical Engineer with the Canadian Niagara Power Company, where he was instrumental in placing in operation the first 10,000-hp turbines ever built. In 1916 he was granted his degree in Mechanical and Electrical Engineering from the University of Toronto, and in 1924—the year he resigned from Ontario Hydro to enter private consulting practice—the University awarded him a Doctorate in Science, *honoris causa*. His greatest single undertaking in private practice was the Shipshaw Power Development of the Aluminum Company of Canada at Shipshaw, Quebec. Completed in 1943 with a total installed capacity of 1,200,000-hp, it was at that time the largest hydro-electric installation housed under one roof. Acres died in 1945 in his sixty-sixth year, after an outstanding career in electrical engineering spanning a period of 42 years, close to half of them spent with Ontario Hydro. His record of achievements as director of design and construction in many parts of the world included the completion of 370 transmission lines, 20 storage dams, and 28 hydro-electric plants with a combined total installed capacity of 2,200,000-hp.

Hand in hand with the prolonged study of the hydraulic factors, engineering surveys were conducted to determine the best location for the canal. In that connection many other factors were involved. Whatever the route to be finally chosen, the canal would have to be built through a highly developed farming area traversed by a complex of railroads, highways and secondary roads which would have to be relocated or for which bridges would have to be provided. An equally important factor was the determination of the sub-surface conditions along the several proposed routes.

Geologically, few areas of the world's surface have been more exhaustively studied than that which lies between the Niagara and Detroit Rivers, partly because of its geological interest and partly because of borings for petroleum and gas made in its paleozoic formations during the last half of the nineteenth century. Despite the great body of evidence available, additional borings were undertaken to check and counter-check the quality of the various construction sites.

When the preliminary plans had matured, it was found that the projected work presented problems comparable in scope and difficulty with those encountered in the building of the Panama Canal, until then the world's most impressive engineering accomplishment. The material to be excavated alone totalled 17,000,000 cubic yards, or five times the volume of the Pyramid of Cheops, and the concrete to be poured would amount to 450,000 cubic yards. Before construction could begin, 82½ miles of standard-gauge electric railroad had to be laid and units of unheard-of design and capacity had to be developed by the numerous construction-equipment manufacturers engaged by the Commission for this work. Among such units were 8-yard electric shovels with 90-foot booms capable of loading a 20-ton dump car on a track 60 feet above them in one minute and a half. From the beginning it was decided that the actual construction work would be undertaken by the Commission instead of being let by contract, and many special pieces of equipment were designed and built by Ontario Hydro's engineers.

Once work was actually underway engineers and construction men had to cope with unexpected problems: the wartime shortage of men and materials; unforeseen excavation difficulties despite the innumerable borings; the need for larger and larger equipment as the job advanced and plans were progressively enlarged. At the peak of construction, 10,000 workers were employed with a semi-monthly payroll of $750,000, all of which had to be paid in cash. The chief accountant and paymaster, G. A. Honsberger, recalls having to travel the line every fortnight in a pay car protected by armed guards.

From May 1917 to December four and a half years later the Niagara area resounded with the tumult of construction. Miracles became commonplace, and seemingly impossible problems found solutions in imaginative expedients never before attempted. But as the job progressed, so did certain anxieties connected with it, principally the mounting costs. Part of those costs could be readily explained by the increased costs of material and labor, by the considerable enlargement of the initial project, and by unforeseen difficulties which occur on every big construction job and which no amount of planning can forestall. Periodically, however, Beck and his fellow-commissioners found it necessary to make public the revised estimates, but these varied so greatly from time to time that both the Government and the public first became puzzled, then alarmed. Strengthening these reactions were the criticisms of various aspects of the project made by a segment of the press on grounds of wartime patriotism, by disinterested experts, and by Hydro's natural enemies.

As the person most concerned politically, and in a sense financially, Premier Hearst felt the situation sufficiently grave to suggest that the Commission call in eminent outside engineers to give independent opinions on the work of Gaby and Acres. By the time consultants were finally called in, however, a new premier was in office, E. C. Drury of the United Farmers of Ontario. The first consultant, whom Beck himself nominated, was the famous American dam-builder, Hugh L. Cooper. His two reports, rendered in August and October of 1920, were sharply critical. Cooper recommended several changes of design, affecting the structure of the floor and walls of the canal and the system of intake tubes devised by Ontario Hydro's engineers and their consultant, R. D. Johnston. Cooper supplied various estimates on alternative procedures, with costs ranging from about 61 million to 89 million dollars, and a completion date for five or six units by January 1, 1923.

Dissatisfied with these findings, Beck promptly procured the appointment of four other experts: the American engineers F. L. Stuart and H. S. Kerbaugh, and R. S. and W. S. Lea, two Montrealers who had designed many hydro-electric installations in Canada and elsewhere. Stuart and Kerbaugh approved of the Ontario Hydro engineers' design and believed that the Queenston-Chippawa GS could be in operation by September 1921, but only on condition that additional excavating and other equipment be provided. The recommendation was accepted.

The Montreal engineers were not so approving of Ontario Hydro's design as were Stuart and Kerbaugh, but neither were they as disapproving as Colonel Cooper. They reported that on the whole Ontario Hydro's engineers had done a competent job and that the great increase in costs was not due to engineering inability or miscalculation.

The report of the consulting engineers was made in the autumn of

1920 and with labor again more readily available, though at double the pre-war wage scales, the great work moved toward its first production of power at the end of 1921. Rock crushers were installed; cement plants established at Walkerton and Durham; and large areas flood-lighted to enable day and night shifts to work continuously.

By November of 1921 the project was close enough to completion for Beck to invite a group of newspapermen to inspect the Canal. In a statement issued on this occasion the Commission pointed out that during the fall of 1920 and the first half of 1921 high labor turnover had badly retarded the progress of construction. However, despite these and other difficulties, concluded the statement, "Two of the five units are practically installed, or 110,000-hp, and will be ready for operation in December of this year and January 1922. The maximum capacity of the canal will be from 550,000 to 600,000-hp, and the maximum installation in the generating station will be five units of 55,000-hp each and five units of 75,000-hp each, or a total of 650,000-hp."

A month later the plant was formally opened and, in January 1922, as promised, the first unit was put in operation. The powerhouse was being built in sections, to accommodate successive installations of generating units, the ninth and last of which was to be in service in December 1925. The total cost of the completed plant was now estimated at $76,000,000.

The inaugural ceremony on December 29, 1921, was attended by representatives of the Government, the coöperating municipalities, and the Ontario Municipal Electric Association. As the turbine gates opened and the great generators began to hum, Premier Drury switched on a huge illuminated sign reading: THE LARGEST HYDRO-ELECTRIC PLANT IN THE WORLD. It was indeed a great victory for Sir Adam and his engineers; a victory dimmed only by the nagging testimony of the accountant's figures. They showed, beyond any possibility of contradiction or evasion, that for one reason or another Queenston-Chippawa had cost a great deal more than anybody had expected.

The turbines in the Queenston-Chippawa powerhouse had barely begun to turn when Premier Drury wrote to Beck expressing uneasiness about the spread between the estimated and the actual cost of the project. According to Colonel Carmichael, in less than a year costs had exceeded estimates by about 20 million dollars. It was time, suggested the Premier, for the Chairman to stop making one-man decisions and begin admitting the other two commissioners to his counsel. They, and the Government, must have a hand in deciding what was to be done about the unforeseen costs of the Queenston-Chippawa development, and of the Nipigon plants.

What the Premier obviously had in mind was an inquiry by a Royal

Commission, and despite the protests of Beck and his OMEA supporters, such an inquiry was duly ordered.

As Chairman, Drury appointed an able Toronto lawyer, W. D. Gregory, in politics a Liberal. The other members of the Hydro-Electric Inquiry Commission were: J. Allan Ross, a Toronto business-man, R. A. Ross, a Montreal engineer, Lloyd Harris, a manufacturer, and M. J. Haney, a civil engineer and former banker.

The directives for the inquiry did not limit the Commission to the investigation of the spread between the engineers' estimates and the actual cost of the Queenston-Chippawa development: they included a mandate to report upon "any other power developments undertaken by the Hydro-Electric Power Commission and generally all matters of expenditure and administration of the said Commission . . . and to make such suggestions and recommendations in connection with or arising out of the subjects thus indicated as may be desirable, and to report upon the evidence and facts brought out by the investigation, along with such findings."

While the Commission was still at work, the National Electric Light Association, propaganda arm of the American private-power interests, launched another attack on Hydro. This time NELA's spokesman was William S. Murray, an engineer with a distinguished record. He had conducted the Super-Power Survey of the Atlantic Seaboard between Boston and Washington for the United States Government. He had also been employed in Canada as a consultant. His partner, Henry Flood, Jr., had been Secretary-Engineer of the Super-Power Survey organization.

Murray's "impartial study" appeared in the form of a 223-page book packed with maps, tables and charts. Beck considered it sufficiently formidable as propaganda to require a detailed reply, which he duly issued. Matching table with table and chart for chart, Ontario Hydro's indefatigable Chairman demolished Murray's *ad hoc* study point by point. In a personal foreword to this pamphlet, Beck concluded that:

"So long as the 300-odd municipalities including practically all of the cities, towns and large centers now coöperating in the Province of Ontario for the purpose of supplying the people with electrical ser-vice 'at cost' retain their confidence toward each other and toward their Commission, no assaults, no matter what their character may be, can prevail against their great and successful coöperative undertaking."

Between November of 1922 and the spring of 1923 the Gregory Royal Commission submitted four interim reports to the Legislature. They dealt with the Thunder Bay System, the Central Ontario System, the Essex County radials and the Guelph radial. Each was shown to have an operating loss, contributed to in some cases by the accumulation of renewal reserves. The causes of the losses were examined and means to cope with them were recommended. Beck's use of general Hydro funds to finance radials was criticized.

Publication of further interim reports by the Gregory Commission was postponed by Premier Drury's announcement that the Legislature would be dissolved in May, and an election held at the end of June. In the event of a defeat of the hostile Drury Government, Beck and his supporters saw a chance to quash the Gregory Commission. Accordingly, although he had hitherto talked much about the importance of "keeping Hydro out of politics," Beck now allowed himself to be named at the Conservative convention in London as a candidate for the Legislature. Ontario's legislators, he declared in his speech of acceptance, had been misled about Hydro; this would scarcely have happened if he had been on the floor of the House. The Gregory Commission had gathered information to be used by Hydro's enemies.

"These Commissions," the Chairman asserted, "have cost enormously, through the inefficiency they entailed and the persecution of honest, decent men in our employ . . . hounded from pillar to post by the Inquiry Commission, biased and prejudiced and unfair in its deliberations."

As expected, the electors retired Drury's UFO-Labor coalition and put in office a Conservative government headed by G. Howard Ferguson, who, as it happened, rather disliked Ontario Hydro's Chairman. Beck himself was elected by a majority of almost two to one over the combined votes cast for his Liberal, Labor and Independent opponents.

Attempts at the next session of the Legislature to have the Gregory Commission dismissed resulted only in a belated and ineffective effort by Premier Ferguson to limit the investigation to the Queenston-Chippawa development. The final report of the Commission covered the entire range of reference specified by the Drury Government and was presented at the 1924 session of the House. It contained many volumes of evidence, findings running to 65,000 words, and a summary of 5,000 words. The full report was not published, a circumstance made much of by Hydro's enemies, although the findings and summary were adequate and uncensored, and were widely quoted in the press.

The Gregory Commission approved the principle of public ownership of water power and its exploitation by and for the people. It praised Ontario Hydro's engineering department and described Ontario Hydro's plant as "exceptionally well operated." Ontario Hydro's accounting system, it reported, was adequate, and the organization administered by Ontario Hydro was financially sound.

The report became critical with respect to Ontario Hydro's tendency to defer the refunding of its obligations, and its submission to the gov‹ ernment of cost estimates so low that it was hard to believe they could have been submitted as representing the probable costs. For example, the original estimate for the Queenston-Chippawa project had been $24,316,815 for a 300,000-hp development. Actual cost of the completed plant, with a capacity of 560,000-hp, would be about $84,000,-000, as then estimated.

The Gregory Commission Report also condemned the use of part of the renewal funds for the payment of balances due by municipalities, as well as the use of $1,100,000 of Ontario Hydro's funds for radial railway purposes. Ontario Hydro had made a big mistake, said the Report, in becoming involved in the radial projects, and the sooner it pulled out the better.

In general, the Report condemned any use of Ontario Hydro's funds for unauthorized purposes. On one occasion, it declared, Chairman Beck, "having good reason to believe that payment of certain accounts would not be authorized by the Commission, if submitted to it . . . had cheques for payment of them issued without the sanction or even the knowledge of the Board and without there being any legal authority for payment."

Sir Adam was praised for building up Hydro, for defending it against unscrupulous attacks, and for organizing an able staff, free of incompetent placemen. He was reprimanded for his lack of frankness and consideration for others, for his high-handed dealings with the Government and with his colleagues, and for his deliberate misleading of the Government and the municipalities through the submission of estimates which he knew to be inadequate or unsound. He had submitted as total estimates figures that his engineers had perhaps meant as partial estimates, simply to get what he wanted. In some cases Beck had waited three years before telling the Government that costs had been increased on a given project.

"No head of any department of the Government," declared the report, "doing the things which he (Adam Beck) did, would have kept his position." He had stayed in office simply because he had "created a political force which governments as a rule are unwilling to antagonize. . . . The spectacle of the Hydro, or rather the Chairman, exceeding legislative appropriations by millions, spending for one purpose millions entrusted to it for other purposes, making agreements clearly beyond the powers of the Hydro while auditors protest and governments and legislature look on, has its humorous as well as its serious side."

The report recommended "complete revision and probably consolidation of the Power Commission Act and of other statutes affecting the powers and duties of the Commission." The Legislature should consider "the question of the adequte taxation of the properties of the Commission and the fixing of a reasonable price for water rental." By 1923, the report noted, the Provincial Government had advanced 162 millions for Hydro purposes, of which only two and a half millions had by that time been returned. Over 55 per cent of the Provincial debt was on account of Hydro.

The report further recommended that the Central Ontario System be "unscrambled" and its operation placed under the terms of the Power Commission Act. It deplored the rate differential that favored the Niagara System, with the effect of draining small industries from other parts of Ontario into the Niagara area. This differential, it found, was due not only to the large consumption of power within a small area, but to the action of the Commission in making the sinking fund repayments over a period of forty years at Niagara, but compressing them into thirty years elsewhere.

These criticisms were balanced in some degree by the striking tribute paid to Ontario Hydro's engineers. The Gregory Commission was of the opinion that, in view of the magnitude and novelty of the project, Ontario Hydro should have employed competent consultants from the beginning. It found, however, that "the design of the Queenston-Chippawa development was based on the most intricate calculations known in the theory of hydraulics, but even so there was some doubt as to whether or not the result sought for would be obtained through it. It now appears clear that the engineers of the Commission, as designers of this great work, surpassed even their own expectations. The canal was designed to pass 15,000 cubic feet of water per second, but we are advised by our consulting engineer that it is capable of passing 18,000 cubic feet of water per second or more. The engineers stated that they hoped to get 30 horse-power per second foot, but the test which we have made indicates that it would develop 500,000 electrical horse-power, but it seems clear that it will, on completion, develop 550,000 electrical horsepower, a most substantial increase. The plant now has an efficiency of over 90 per cent, an unusually high figure and one which indicates a fineness of design seldom, if ever, attained in a work of this character. It is, in short, a magnificent piece of engineering."

It was unfortunate, said the report, that the building of Queenston-Chippawa coincided with a time of peak prices; costs had been further increased by rush-schedule operations. But most of the many charges of irregularities and wastefulness were baseless. "Concerning the management," concluded the report, "there is not a breath of suspicion of any personal wrong-doing."

Many of the Gregory Commission's well-considered recommendations were subsequently adopted. The report had cost half a million dollars and probably was worth at least that much to the Government, the municipalities, and the people of Ontario. But it did not please Adam Beck. The Premier, when tabling the report in the House, had remarked how gratifying it must be to Ontario Hydro's Chairman that the Gregory Commission had "completely vindicated the Hydro project." Later, however, Ontario Hydro's Chairman spoke for several hours, passionately defending himself against certain of the charges in the report, particularly the alleged misapplication of Hydro funds to the use of the radials.

Beck quoted from statements the Auditor had made in a report to the Government on March 19, 1920, to the effect that Sir William Hearst had promised a Provincial guarantee for the bonds covering that $1,100,000 expenditure. Only $550,000 of it had been spent on radials; the balance on supplying evidence to the Sutherland Commission, which had reported adversely on the radial projects.

Clearly Beck had not forgiven the Drury Government for setting up the Gregory Commission. "The insinuations of the opposition," he said, "have been thick as flies for three long years—it seemed more like thirty years—and if the Drury Government had remained in office much longer there would have been no Hydro except as a government department debauched in politics."

At which point Beck sat down to give the Legislature a chance to register its confidence in Ontario Hydro and Hydro's Chairman. It did so by approving Ontario Hydro estimates for the year amounting to $22,000,000.

CHAIRMEN OF ONTARIO HYDRO

The dynamic drive and unflagging devotion of its first Chairman, Sir Adam Beck, piloted Hydro through its turbulent early years. More than any other factor, Beck's undeviating sense of purpose ensured the success of the enterprise. While Sir Adam Beck lived, Hydro was Beck and Beck was Hydro.

C. A. MAGRATH
1925-1931

J. R. COOKE
1931-1934

T. S. LYON
1934-1937

T. H. HOGG
1937-1947

From those early days onward, Hydro has been fortunate in having the services of men of outstanding ability and integrity, devoted to the organization and to the principles that brought it into being. The challenge of its beginning has been carried forward and given new impetus by each successive Chairman—each has contributed qualities that served to consolidate Hydro's phenomenal success.

R. H. SAUNDERS

1948-1955

R. L. HEARN

1955-1956

J. S. DUNCAN

1956-

THE RADIALS DISPUTE

15

Adam Beck spent over a decade, at the height of his career, fighting for electrical railways, to be built with the help of provincial subsidies, to link city with city and rural villages with their urban markets. Radials, he insisted, were a natural and essential corollary to the successful, rapidly expanding system of publicly owned and operated utilities of which Ontario Hydro was the heart. Beck first projected the vision before an audience of his supporters at Brantford on October 22, 1912.

What Beck actually envisaged was a system similar to those he had seen operating successfully in the highly developed industrial regions of Europe and parts of the United States: a network of high-speed electric railways radiating from Toronto and connecting with all the surrounding municipalities. With the deepening of the St. Lawrence Seaway, Toronto, he believed, might well supplant Montreal as the Atlantic gateway to the heartland of Canada. And, not too incidentally, the development of radials would supply a vast new market for Hydro power.

Late in 1912, in Toronto, Beck addressed a meeting of OMEA, with G. R. Geary in the chair. The Ontario Hyhro staff, he said, would estimate the cost of radials, but it would be for the Government to finance them. Like Hydro, the radials should be operated on a service-at-cost basis. The meeting resolved: "That . . . it is desirable that a system of electric railways, including street railways to be owned by the municipalities, be established and built; and further, that the Hydro-Electric Power Commission be requested to look into the advisability and practicability of building such a system, and to furnish a report thereon to this Association."

In April 1913 Beck introduced in the Legislature a bill granting authority to the municipalities, subject to government approval, to build and operate hydro-electric radials at their own expense. Passed

137

in May, the Hydro-Electric Railway Act was purely permissive and embodied no financing provisions so far as the Province was concerned. It did not, however, exclude the possibility of a Federal subsidy which Beck had adumbrated in some of his speeches.

Finally, at a meeting of municipal delegates arranged in Toronto in December 1913, the Ontario Hydro Chairman proclaimed that only by building a radial network, and building it now, could the alarming increase of emigration from farm to city be halted before it was too late. No doubt by prearrangement, the delegates responded by forming the Hydro-Electric Radial Union of Western Ontario and passing a resolution asking Ontario Hydro to build lines wherever needed.

The new pressure group and the resolution were opening shots in a campaign which was to continue for eleven years and embroil the entire province in bitter, acrimonious debate. During the months following the Toronto meeting in December, Beck continued his agitation in other cities of the province. Another Hydro Radial Union was organized in the Niagara district, and yet another pressure group, the Great Waterway Union, was established to add to the growing force of propaganda. This propaganda, by the winter of 1914, had acquired a three-fold objective: Hydro radials, the diversion of more water at Niagara to justify the Queenston-Chippawa project, and the development of the St. Lawrence River for power and navigation.

Beck's proposal was hailed by his loyal followers, and particularly by those in the rural districts who knew how bad roads could be, as a practical broad-visioned solution of what had been regarded as an insoluble transportation problem. Furthermore, the people of Ontario along with those of the rest of the country were in a railroad-building mood and ready for such a solution. Since the turn of the century, Canada had enjoyed the greatest influx of immigrants in her history; with it had come the construction of not one, but two new transcontinental railroad systems—the federally backed Grand Trunk Pacific, and Mackenzie and Mann's Canadian Northern. Meanwhile, the mileage of electric railways had increased from 675 miles in 1901 to 1,047 in 1910 and 1,308 in 1912.

In February 1914 Beck, who had been in England reportedly considering an appointment to the office of Canadian High Commissioner, returned to Toronto to address a meeting of the Associated Boards of Trade. In his speech he made it clear that he did not intend to leave Hydro and that he did intend to press the drive for radials. A fortnight later, at a meeting of 700 municipal delegates in Stratford, he confided his dream of a Greater Hydro which would control and operate not only radials but also telephone and telegraph services, gas and water systems, and other publicly owned utilities. All this would be possible, he declared, if only Ontario Hydro were permitted to generate its own power at Queenston, deepen the St. Lawrence Seaway, and thus open the Canadian heartland to ocean-going ships. "Nothing is visionary," declaimed Ontario Hydro's Chairman, "with regard to this wonderful energy"—the "white coal" with which Canada could now compete successfully with the black coal of the United States.

With Ontario's permissive Hydro-Electric Railway Act already enacted, financing became the primary consideration, and on March 25, 1914, there converged on the Federal House of Parliament in Ottawa a delegation, two thousand strong, with memoranda urging the Government to take immediate action to finance Hydro radials, and to open negotiations with the United States respecting the Niagara diversion and the joint development of the St. Lawrence.

The descent on Ottawa was at best a dubious success. Prime Minister Borden received the delegation graciously, extended the Government's assurance of approval of the Hydro-Electric movement, but took no steps to implement its demands. A week later, however, another deputation consisting of 50 prominent leaders of the Hydro Radial Union besieged the Government of Ontario in Toronto. It asked the Government that Hydro be permitted—and provided with funds—to develop power at Niagara and elsewhere in the province, that legislation be passed guaranteeing bonds for construction of Hydro radials, and authorizing a 50-year period for the bonds to mature, with no sinking fund payments for ten years so as to give the radials a fair start.

In view of the approaching elections, both political parties adopted conciliatory tactics. A Government bill, the Hydro-Electric Railway Act, 1914, passed with Opposition support, repealed the 1913 Act which had required the municipalities to bear all costs of radial construction. The new Act gave Ontario Hydro "power to act on behalf of the municipal corporations interested, to issue bonds for carrying on the work, taking the debentures of the corporations as security, and thus providing ways and means for the financial undertaking of the work." A year later the Act was amended to enable part of a township to bear its proportion of the construction and expense of radial railways, and to permit the purchase of existing lines.

Only the outbreak of war prevented full and rapid implementation of the large-scale radial program authorized under these broadened provisions. All of Ontario Hydro's projects were in fact affected, but the increasing demand for power acted to spur the Queenston-Chippawa development, while radial projects were given low priorities.

During the first year of the war, Beck continued to push the drive for radials so effectively that when the by-law election was held in the districts east of Toronto the project was approved by large majorities. A few months later, when four private radial railways asked the Legislature for franchises, Beck successfully opposed them, on the ground that in radials, as in electric power, Hydro needed to have a province-wide monopoly.

In February 1915 Toronto's special committee on rapid transit, at Beck's urging, petitioned the Legislature not to grant new radial charters to private interests. Toronto, said Beck, would be the logical hub of the entire publicly owned radial system which he envisaged. Though he agreed in a subsequent statement not to press for radials during the war, Beck at this point demanded that both the Provincial and the Dominion Governments declare their intentions. Various Hydro groups promptly backed this demand. A new propaganda organization was formed—the Ontario Hydro-Electric Railway Association. Its announced objectives were: to plan and promote radials; to see that necessary legislation and government aid for radials was forthcoming; to help municipalities to carry radial by-laws; and to prevent private interests from getting further radial charters.

Fifteen hundred Hydro supporters, marching to the skirl of bagpipes, invaded Queen's Park on March 26 and presented a resolution asking the Province to provide a subsidy of $3,500 a mile for municipal Hydro radials. Premier Hearst temporized. He needed time, he said, to study the liabilities which this proposal involved for the Province.

In July the London and Port Stanley Railway, now a utility of the City of London, staged a gala celebration to demonstrate its newly electrified service. Unhappily there was a breakdown on the outward journey, south of St. Thomas, and Sir Adam joined Fred Gaby and other Ontario Hydro engineers in attempting to repair a broken pantograph-trolley. Finally the train coasted to Port Stanley, the picnic was held, and a steam locomotive towed the train back to London.

The automotive revolution was greatly accelerated during the war, largely because the motor truck so successfully met the requirements of wartime transport. In 1918 in the United States, when the railroads could no longer cope with the increased traffic, R. D. Chapin, Chief of Highway Transportation, loaded 30,000 trucks with war supplies and drove them from the middle west to the Atlantic seaboard. Beck and his supporters read of this feat, but failed to grasp its implication in the drive for radials, which by then called for a 2,000-mile interurban network to cost $90,000,000. Don't worry, Beck told his audiences. Radials would be a profitable investment from the beginning.

The Ontario Legislature had already chosen to worry. A clause in the amended Hydro-Electric Railway Act, passed in April 1916, prohibited action until the end of the war.

Before then, Toronto worried about the problem of providing a suitable entrance into the city for the projected radials. A committee was appointed consisting of three experts: the City's Works Commissioner, R. C. Harris, the engineer of the Harbor Commission, E. L. Cousins, and Ontario Hydro's Chief Engineer, F. A. Gaby. Their report recommended the construction of a system of electric railways, terminals and yards; it advised the purchase of the Toronto Street Railway Company when its franchise expired in 1921, and the creation of a new municipal administrative authority, the Toronto Transportation Commission, to be charged with the control of the City's railways and buses.

Presumably because of the potential conflict of authority with Ontario Hydro, Beck objected violently to these recommendations and denounced Gaby for agreeing to them. When Lionel Clark, Chairman of the Harbor Commission, attempted to defend the engineers, he and Beck almost came to blows.

To ensure passage of the radials by-law at the 1916 New Year's election, Beck mobilized all his political forces: the Hydro-Electric Railway Association, the Ontario Municipal Electric Association, the pro-Beck faction in the City Council led by Mayor Church, the Orange Lodges (for which the Toronto *Telegram* was the spokesman) and, unofficially, the ward organization of the Conservative Party.

The Toronto Board of Trade, once a staunch supporter of Beck, noted that Toronto, with only one vote out of thirty-one—the number of participating municipalities—would be responsible for one-third of the cost of the proposed radial system. The Board was also concerned about surrendering control of the City's streets to Ontario Hydro, the duplication of steam and electric lines, and the possibility that the system might soon become obsolete because of the growth of automobile trucking. For these reasons the Board voted to oppose the radials by-law.

The day before the election Beck released a letter to Mayor Church promising to support any legislation that might be required to prevent unauthorized interference with the city's street system. Probably the letter had some effect in achieving the wide majority—21,161 for, 5,766 against—by which Toronto approved the by-law. London, Beck's home town, came close to voting against it, but the count of the municipalities was 23 in favor to only three opposed.

Passage of the by-law gave the radials a right-of-way along the Toronto waterfront. But since the width of the right-of-way was not specified, the Toronto Board of Control foresaw another battle. Shortly after the election it advised the Municipal Council to require that Beck make good on his pre-election promise with respect to safeguarding the City's control of its streets.

Beck's only reply was to declare, in an address to the Hydro Radial Association, that Ontario Hydro would go ahead with radials, prepare the plans and specifications, buy the necessary rights-of-way, and construct approved engineering works; this in the face of the country's tightening shortage of men, materials and money. The Canadian Expeditionary Force needed replacements, and the Dominion Parliament was asked to vote conscription measures which were bitterly opposed by the Province of Quebec and by the farmers all across Canada. Since neither the Liberals nor the Conservatives would risk the unpopular act of imposing military service, Premier Borden decided that a Union government was necessary.

Meanwhile the Western Front bent and wavered under the German offensive of 1918. Responding to a desperate appeal for help, the Canadian Parliament cancelled the exemptions that had been granted to young farmers.

Political revolt, long brewing, came to a head when five thousand farmers from Quebec, the Ontario agricultural districts, and the prairie provinces invaded Ottawa. Conscription, they shouted, was a breach of a solemn election pledge, and it would reduce farm production by at least a quarter.

The Premier cut them off. His overriding loyalty, he said, was to the fighting men in France. If he let them down, and the Germans reached the Channel ports, there would be no further need for farm or any other war-essential produce. The order ending exemptions would stand.

The frustrated farmers of Ontario were not through, however. That summer the tide turned on the Western Front. When the war ended in November the Provincial Government of Sir William Hearst faced imminent dissolution. In 1919 there would be a general election. The farmers, organized in a new political party—the United Farmers of Ontario—were determined to win it.

As before, Beck held the balance of power. The farmers considered him not only as a big man, but their man—more so, certainly, than Sir William Hearst. Possibly, thought the Conservatives, they would stand a better chance of being returned to office if they ran Beck for Premier instead of Sir William. But would Beck be a candidate? He would, he declared, but only on the condition that he was given a free hand with

Hydro and the radials; otherwise he would withdraw from politics altogether. In the end Beck decided to run, but as an Independent.

The officers of the UFO were R. H. Halbert, President; E. C. Drury, Vice President; and J. J. Morrison, Secretary and organizer. Both Halbert and Morrison had been violently against conscription, whereas Drury had fought the 1917 federal election, unsuccessfully, as a Liberal conscriptionist. Of the three he was the most politically acceptable, as well as the best equipped with respect to his grasp of economic problems. None of the trio was an enthusiastic admirer of Beck or of his policies with respect to radials.

As an independent candidate, Beck campaigned almost solely on the issue of radials. In March he persuaded the City of Hamilton to reverse its 1918 decision and permit the construction of the Toronto-Niagara radial, so that now Ontario Hydro was enabled by Order-in-Council to go ahead with construction and issue bonds to provide the money; construction of the Kitchener-London line was also begun. In May, people living east of Toronto petitioned Ottawa to incorporate the Toronto Eastern Railway into the Hydro radial network. Beck envisaged a continuous line running around the crescent shore of Lake Ontario from Bowmanville via Toronto, Port Credit and Hamilton to St. Catharines and Niagara, and one from Hamilton to Guelph and Elmira. The total cost, he estimated, would be $30,000,000. According to Ontario Hydro's 1919 report, construction was then already started on the 43 miles of line between Bowmanville and Toronto, and 14 miles of rail from Bowmanville toward Whitby were completed. The cost was estimated at eight million dollars. Revenue and operating expenses were estimated and balanced with a comfortable margin of solvency. In short, the realization of Beck's grandiose radial project seemed on the way.

At that time the fastest trains between Toronto and Niagara Falls took two and three-quarter hours; American experts predicted that if radials could average only 40 or 50 miles an hour, they would be profitable. But even in 1919 the widely-expanding automotive industry was beginning to undermine the premises of these optimistic estimates. Already the fruit growers on the Niagara peninsula were turning to motor vehicles for farm-to-market hauling.

Beck's opponent in the October election was Dr. Stevenson, an esteemed physician with little political experience. Beck probably would have won handily if the Conservatives, irked by his bolt from the Party, had not chosen to teach him a lesson. When the votes were counted, Beck was defeated by a small margin, but so were the Conservatives, including the Premier, Beck's fellow-commissioner, I. B. Lucas, and four out of five of the former Conservative members of the House. The Liberals were also beaten. The vote stood: UFO 45 members, Liberals 29, Conservatives 25, Labor 11, Independent 1—but that one was not Beck.

The UFO was in. A new party, without legislative or administrative experience, was asked to form a government. Whom would it choose for the premiership?

It might well have turned to the defeated Independent candidate if Sir Adam had played his political cards more discreetly. The offer was made and Beck accepted—again with the proviso that he be permitted to do what he thought best for Hydro and the radials. Beck's "boys" were jubilant and a Hydro meeting at Windsor went so far as to pass a resolution urging that he be made Premier.

That was too much for the UFO leaders. Were they to have their policies decided by Beck's pressure groups? All of them were dubious about radials, Morrison in particular. After some conferring the premiership was offered to Drury, who accepted and chose a scratch cabinet of eleven. Lieutenant Colonel D. Carmichael, DSO, MC, was appointed to take the place on the Commission left vacant by the death of W. K. McNaught.

Beck was offered an appointment as Drury's Minister of Power, which he declined, as he had rejected similar offers from Whitney and Hearst. His concern was now to ensure continuity in his office of Chairman. With their usual responsiveness, several hundred delegates from the Hydro municipalities now called on the Government to fix the term of Ontario Hydro's chairman at six years; they further requested that in the future there be five commissioners instead of three and that two of these be nominated by the municipalities.

The UFO leaders felt obliged to receive such suggestions respectfully. But they didn't like Beck and they were worried about his radials. At the UFO annual convention the farmers had passed a resolution which cited sobering reasons for viewing the whole radial program with alarm: the enormous cost, the duplication of existing steam lines, the expanding facilities for road traffic. . . . But the UFO, which needed the entire Labor bloc to give it a majority of one vote in the House, was not in a position to fight Beck openly. Labor liked Beck and favored the radials. So did many Conservative and Liberal members.

The UFO paper, the *Farmer's Sun*, felt less inhibited by these realistic political considerations than did Drury and his governmental associates. At any rate, during the winter and spring of 1920, while Beck was ill in England, the paper denounced the radial plan, pointing out that in the United States 90 per cent of the electric railways were losing money. More of the Hydro municipalities, it reported, were showing deficits than surpluses. And how had it happened that the construction costs of Queenston-Chippawa reached four times the estimates?

Beck replied to these attacks on his return from England. The original Queenston-Chippawa estimates, he pointed out, were based on the much smaller development then contemplated. The United States radials were losing money only because they had to pay the high power rates charged by private companies. He intended to go ahead with the radials, for which the people of Ontario had voted, but with a limited program.

This was the first time that Beck had spoken of limiting his radial program and clearly the limits were not of his choosing. During the spring of 1920, the Drury Government started to apply the brakes, in the form of an investigation on financial relations between the Province and the Commission made by the independent auditor, George T. Clarkson, in whose professional competence Beck had publicly expressed confidence. At the same time a committee of Parliament studied the advisability of a flat rate for power.

Beck, too, by way of strengthening the case for his "limited program," employed an independent investigator. W. S. Murray, a New York consulting engineer, was asked to report on certain aspects of the radial scheme. Murray's report, made public early that summer, dealt with the continuous line from Bowmanville to Niagara and the line from Hamilton to Guelph. While confirming Beck's estimates of costs and probable revenue, Murray advised that construction be delayed because of the current high costs. Meanwhile he recommended that the Commission proceed with its financing plans.

With the backing of this report, Beck now submitted to the Federal Minister of Railways a proposal to buy the Toronto-Eastern, the Toronto Suburban, and the Niagara-Toronto lines at a cost of nearly $7,000,000, using for that purpose provincially guaranteed 4½ per cent 50-year bonds. The Province, as Beck took care to remind the Drury Government, had already spent or agreed to spend $19,000,000 for the purchase of rights-of-way and for constructing and operating various lines, including the Sandwich, Amherstburg and Essex Railway, the Toronto and Eastern Railway, and the Port Credit-St. Catharines line.

The Drury Government scarcely needed this reminder. It had the Clarkson report, which found that about 50 per cent of the Province's debt of $100,000,000 was on account of advances of cash and securities, $40,000,000 to Ontario Hydro and $10,000,000 for the purchase of the Central Ontario System now being operated by Ontario Hydro. About $29,000,000 would be needed by Ontario Hydro during the next two years, together with guarantees of between $25,000,000 and $26,000,000 for Hydro radials.

In view of such costs, was the Province justified in making the commitments that Beck was demanding? Should it guarantee bonds for the construction of the Toronto and Eastern radial, as OMEA and the Oshawa Board of Trade were currently urging, against the violent protests of the *Farmer's Sun* and the Ontario Hydro Information Association, recently organized by UFO leaders?

To resolve the radials controversy, Drury resorted to that invaluable device, the Royal Commission. Its Chairman, Justice Sutherland, had been a member of another Royal Commission which during the war had investigated the use of Niagara water by the Electrical Development Company. Its members, carefully chosen to balance the conflicting political forces with independent experts of established reputation, were W. A. Amos, Vice President of the UFO, Fred Bancroft, a Toronto Labor leader, A. F. MacCallum, City Engineer of Ottawa, and C. H. Mitchell, Dean of the School of Science of the University of Toronto, who had given expert advice at the inception of the Hydro movement.

Beck and his supporters greeted the appointment of the Sutherland Commission with roars of protest. Why could not the question be left to the judgment of the ratepayers, asked the Hydro-Electrical Radial Association. Was the Royal Commission merely a club with which Drury intended to beat the radials scheme to death? The Dominion Trades and Labor Congress implied as much; its President, Tom Moore, declared that organized labor would not sit idly by and let the whole radials project be blocked.

The Royal Commission report, based on testimony obtained from American and Canadian railway and other experts competent to appraise all aspects of the radials question, was indeed enough to bring the whole project to a halt. Radials, said the Commission, were becoming increasingly unprofitable both in Canada and the United States. Two Canadian steam railways covered some of the very same territory to be served by the radial system Beck proposed and road traffic was giving them more and more effective competition; this the Province was encouraging by a $25,000,000 good roads program.

Construction of radials, concluded the majority of the Royal Commission, would not be warranted without authoritative assurance, which was not likely to be forthcoming, that the system would be self-sustaining. Competition with the newly created Canadian National Railways, also publicly owned, was obviously undesirable. The prodigious cost of the radials scheme—an estimated $45,000,000—was more than the Government should undertake, at least until the Queenston-Chippawa development was completed and in operation.

Fred Bancroft, speaking for labor, issued a minority report of one. Disagreeing flatly with the majority view on all points, he predicted that the radials would be self-sustaining and urged their construction by Ontario Hydro, which had demonstrated the financial and economic soundness of public ownership and its capacity to serve the people.

Beck's response to the majority report of the Sutherland Commission was one of outright defiance. In a series of addresses to Hydro groups he urged the municipalities to pay no attention to the findings of the Commission and to go ahead and build as they saw fit. They needed, he insisted, neither advice nor permission from the Government or from any Government-appointed commission.

Actually, if Beck's radials scheme were to materialize, the Government would have to guarantee new bonds. This Drury refused to do. By way of imposing additional curbs on Beck, the Premier altered the composition of the Commission, replacing the Hon. I. B. Lucas with Fred Miller, a member of the Toronto Transportation Commission. He and Commissioner Carmichael could be counted on to vote down the Chairman when it came to promoting the passage of radial by-laws by the municipalities.

However, much of Beck's program was already well advanced. The Sandwich, Windsor and Amherstburg line had been acquired by the municipalities under a Provincial guarantee. Under authority of the Guelph Railway Act of 1921 Ontario Hydro had acquired the Guelph Radial Company for operation. Under a similar guarantee, the Metropolitan Railway and the Port Credit and Scarboro lines had also been acquired. The Toronto Power and Railway Purchase Act of 1921 enabled the City of Toronto to buy the portion of the Toronto and York Radial Railway situated in the city, while the Toronto Radial Railway Act had empowered the city to buy the portion situated outside, and with it the Schomberg and Aurora Railway Company. Finally, the Toronto and Suburban Railway Act of 1922 authorized Ontario Hydro to buy and operate the suburban line on behalf of the city.

Beck's reply to the Sutherland Commission's majority report was issued in the form of a 42-page booklet, which bore his official signature as Chairman of Ontario Hydro. His answer condemned the Sutherland Commission as incompetent—it had been embarrassing for Ontario Hydro to subject its experts to examination by such a tribunal. The radials scheme was sound and must be completed. Surveys, estimates, and working plans were ready for the construction and equipment of 325 miles of radial line. Of this projected system, only 125 miles of new line would have to be built; the remaining 200 miles needed only to be modernized. Completion of the initial scheme would be followed by a province-wide extension of electrified rapid transit lines, economically operated because they would use Ontario Hydro's power at cost.

Beck also declined to consider the Sutherland Commission's alternative recommendation with respect to Toronto, namely that a radial system might succeed locally if it were built as an independent municipal enterprise, rather than as part of the Hydro-sponsored coöperative scheme. Such a system could include the three sections of railway line in or near the city and three others in addition to the street railway system, all about to be acquired by the City.

Beck objected to this proposal on the ground that it would remove the Toronto radial entrances from the control of the coöperating municipalities and make them an integral part of the Toronto Transportation Commission's operations. If this were done, said Beck, it would abrogate the province-wide radials policy of the municipalities and violate the Government undertakings upon which that policy had been based.

By adopting this all-or-nothing position Beck staked too much on the successful issue of his fight in the Toronto City Council and on his ability to swing the Toronto ratepayers into line. The fight in the Council lasted seven days, from August 29 to September 6, 1922. Beck entered the arena flanked by his chief engineer Fred Gaby, the Hydro Secretary, Major Pope, and its counsel, C. C. Robinson. For the Toronto Harbor Board, its Chairman, Home Smith, appeared along with Ex-Mayor R. J. Fleming, Ex-Mayor T. L. Church, and the Board's engineer, E. L. Cousins. The Toronto Transportation Commission was represented by P. W. Ellis and George Wright, with Peter Witt, an American Railway expert, testifying as its consultant. The Board of Trade was represented by its President, D. A. Cameron, along with A. O. Hogg and R. A. Stapells. The Canadian National Exhibition Association was represented by its Vice President, G. T. Irving. The City of Toronto was represented by Mayor Maguire, and aldermen and controllers, together with the City Solicitor, Parks Commissioner and Works Commissioner.

The issue was whether or not Toronto would grant an exclusive six-track right-of-way along the waterfront, with operation of a Bay Street subway by the Hydro radials. Back in 1915 a planning committee composed of Gaby, Cousins and Harris had agreed on a more modest scheme which provided an entrance for the radials through the Exhibition grounds, and a four-track right-of-way—two for the Hydro radials and two for the Toronto Transportation Commission—with city lines and radials and terminals to be under joint control of Ontario Hydro, the City and the Harbor Board. In 1916 the ratepayers had approved the plan. But Beck, who had denounced this reasonable solution of the problem seven years before, was now unwilling even to consider it.

On the last day of the seven-day session the debate was extended interminably. Finally, at five o'clock in the morning, Alderman W. R. Plewman, who had tried earnestly to harmonize the contending interests, got a chance to speak.

"Consider well," said Plewman, addressing Beck and his faction, "before you reject this attempt at mediation. . . . By compromise you get at once a four-track right-of-way. You also get the chance to convince the citizens on January 1 that you must have the use of the additional tracks. Remember this, it would be far better for Hydro's prestige to get a four-track right-of-way by the almost unanimous consent of the Council and with the general approval of the citizens than to force through by a narrow margin an exclusive six-track right-of-way, flout an impressive body of public opinion and impose on Toronto a war to the death between public bodies and leading citizens, all of whom are equally loyal to the community."

The Council debated another hour and then gave the victory to Beck and his supporters by 17 votes to 5. Beck undoubtedly had counted on avoiding another ratifying election by the ratepayers. But in this he had not counted on Drury, who ruled that the controversial 1922 plan adopted by the Toronto Council was so unlike what the ratepayers had approved in 1916 that they must again be consulted.

What ensued during the next four months was Plewman's predicted "war to the death"—it was indeed "the most remarkable civic election in the history of Canadian municipalities." On January 1, 1923, Beck met his Waterloo by a count of 23,129 in favor of the proposal and 28,325 against.

The radials scheme had received a mortal blow. Legislation passed by the Drury Government soon put an end to agreements between Ontario Hydro and the various municipalities concerning the Toronto-Niagara line. The only permitted agreements barred all financial assistance by the Province; over a dozen communities approved these agreements in 1922. But in 1923, when new elections were required, some of the municipalities, including Hamilton, voted against the by-law. Premier Drury authorized building the radial from Toronto as far as

Oakville, and an already granted government guarantee of bonds to the amount of $1,000,000 was confirmed by Order-in-Council; the Toronto-Oakville section, said Drury, was the most likely to succeed.

But time was running out for the radials scheme. In the years that followed, the municipalities that had taken on radials suffered heavy losses. Lines were abandoned and tracks torn up. The Commission carried for many years a heavy radial loan from the Bank of Montreal, secured by provincially guaranteed Hydro bonds, whose worth was maintained by the increasing value of the rights-of-way.

Defeated for almost the first time in his career, Beck was never quite the same man again. He lived just long enough to realize that he, who had been right about so much else, had been tragically wrong about radials.

16

There was an implicit contradiction in the twin slogans Beck had used to win support from the people and from successive provincial governments: "Power at cost" and "Power to the remotest hamlet." Power at cost, yes, but cost to whom? Who was to pay the cost of running expensive rural distribution lines into the back country, to the forgotten townships? The settlers served by these lines? Their number was small and the cost of service to remote farms and hamlets would be prohibitive if the consumers were obliged to pay it all. Yet it was precisely in such places that cheap electricity would effect the greatest social and human benefits. Who, then, would pay for rural electrification? The municipalities, by accepting higher "cost" rates from Ontario Hydro? But if an averaged flat rate were charged everybody, then, in effect, Ontario Hydro's municipal partners would be asked to subsidize the farmers—an invitation which they would probably decline.

Both Ontario Hydro itself and the Provincial Government took steps to resolve this dilemma. Ontario Hydro's Rural Rate Committee studied the problem and in a report dated August 21, 1919, recommended sweeping changes in the policy of supply to rural areas. These recommendations were accepted by the UFO Government and implemented.

By an amendment in 1920 to the Power Commission Act the Province was divided into rural districts, each consisting of about one hundred square miles. These districts were to be operated by Ontario Hydro, each as a unit, with its own cost accounting and rate system. The rates were to be adjusted periodically on a cost basis. Contracts for services were to be signed for a twenty-year period. In every zone, electrical service was to be provided in each of certain specified classes and throughout the whole district at specified rates, based on average conditions in the district. The boundaries of these zones or districts

151

were determined not by township boundaries, but in relation to the economic distance at which service could be supplied from a distribution centre.

At this period the average Ontario farm was from one to two hundred acres in size, which meant that usually a mile of distribution line would service five or six farms. The Commission decided to extend service only on the basis of at least three farm contracts per mile of primary transmission line. This stipulation was later lowered to a minimum of two farms.

The Ontario Hydro recommendations resolved only part of the dilemma. To see what further steps could be taken the UFO Government under Drury set up a legislative committee under the chairmanship of J. G. Lethbridge, a UFO leader. Other members of the Committee were ex-Toronto Controller John O'Neil, a Liberal; W. H. Casselman, UFO; F. H. Greenlaw, UFO; and J. R. Cooke, a Conservative. (Subsequently, Cooke was to become a Commissioner in 1923, and Chairman of Ontario Hydro from 1931 to 1934.)

The Committee released its report in November 1920. Rejecting the idea of a flat rate as impracticable, it recommended, instead, that subsidies for rural electrification and other purposes be obtained through various devices: an overall rental, paid to the Province, or two dollars per horsepower on all power developed in Ontario, to provide about $2,000,000 a year; taxation of Ontario Hydro properties; and transfer of control of all Ontario water from the Federal to the Provincial Government, this to yield another four or five million dollars.

It was estimated that from the total yield of these measures $1,000,000 could be applied to meet annual charges on 12,000 miles of rural lines. As a means of applying subsidies to rural electrification the Committee recommended that small towns and villages be included with Rural Districts, so that each rural municipality or municipal power zone could have a flat rate for all residents within its boundaries.

Passage of the Amendment Act of 1920 was thereupon followed in 1921 by the Rural Hydro-Electric Distribution Act, authorizing grants-in-aid covering up to 50 per cent of the primary transmission lines and cables for delivery of rural power throughout the Province. In 1924 these grants were extended to permit the Province to pay 50 per cent of the capital cost of installing service transformers and meters, and secondary lines on the highway from the nearest distribution centre to the boundaries of the rural customer's properties. The effect of these subsidies was to relieve country districts of half of the investment required for distribution facilities; this amounted to about one-sixth of the total annual cost of power and operation. By two acts in 1930, the

Rural Power District Service Charge Act and the Rural Power District Loans Act, the Government guaranteed a low maximum service charge and offered loans on easy terms to pay for installing wiring and equipment.

These government subsidies were justified as one of several means of promoting agriculture as a basic industry. Other methods were good roads, agricultural colleges and experimental farms. The subsidies were confined to capital investment and guarantee of service charge; each rural power district had to pay its own cost of operation and maintenance, set up reserves for contingencies and renewals, and provide for interest and sinking funds on the investment made by the local authorities.

Ontario Hydro's first rural lines had been built in 1912: the system of provincial subsidies adopted in 1921 resulted in rapid expansion. By 1928 Ontario Hydro was operating about 3,790 miles of rural lines in 122 rural power districts, serving more than 31,000 customers. By the end of 1941 there were 20,000 miles of line representing a capital investment of $39,000,000 and serving 132,000 rural customers.

During the height of the radials controversy, and shortly before the Lethbridge Commission was appointed, an important change occurred in the OMEA—the organization formed in 1912 to represent, and to press for legislation in the interests of, all the municipal utilities that had entered into contracts with Ontario Hydro. In March 1918 the Engineering Section of the OMEA convened in Toronto, and decided to form a separate association, to which end the following resolution was passed: "That since we have been unable to effect a proper organization of Municipalities to consider operation and engineering questions of policy through the formation of the association originally laid down as an engineering branch of the Ontario Municipal Electrical Association, be it resolved that we recommend the establishment of an association of the managers, superintendents and engineers of the different Municipal Electrical Utilities free from the Ontario Municipal Electrical Association and financially independent."

The convention then adopted a constitution and by-laws and elected a slate of officers, comprising president, vice-president, secretary, treasurer, and the chairmen of standing committees. E. V. Buchanan, General Manager of the London PUC, was elected President; E. I. Sifton, General Manager of Hamilton Hydro, Vice President; S. R. A. Clement of Ontario Hydro, Secretary; and R. C. McCollum, Ontario Hydro's municipal accounting expert, Treasurer.

To be known as the Association of Municipal Electrical Engineers (of Ontario), the organization had for its objectives:

1: To further the interests of Municipal Electrical Utilities in Ontario and to foster closer coöperation between the Municipalities and with the parent organizations, viz.: The Hydro-Electric Power Commission of Ontario and the Ontario Municipal Electrical Association.

2: For the mutual assistance of its members, education along technical and commercial lines, and the standardization of methods, apparatus and materials.

The following year, in June 1919, the name of the body was changed to the Association of Municipal Electrical Utilities (AMEU), and it continues to function under that designation as the technological counterpart of the OMEA, constituting a valuable liaison between the engineering staff of the numerous local municipal electrical utilities and Ontario Hydro.

In December 1920, Beck ended his "eighteen-year war" with the Mackenzie syndicate, owners of the Electrical Development Company, which generated power at Niagara; the Toronto Power Company, which transmitted the power to Toronto; and the Toronto Electric Light Company, which distributed the power; and the Toronto Street Railway. As early as 1913 Mackenzie had been willing to sell the railway and the distributing Company for $30,000,000, in a deal engineered by Mayor Hocken. Beck had opposed the purchase, declaring that the price was exorbitant and that the time to negotiate would be in 1921 when the railway company's franchise was due to expire. During the next seven years, offers and counter-offers were periodically exchanged. By 1920 Mackenzie's asking price was $45,000,000. Beck offered $27,000,000. Still outstanding was a bill for $1,000,000, allegedly owed by Ontario Hydro to the Electrical Development Company for extra power supplied to munition plants during the war. Beck offered a half million and finally settled for $700,000. Negotiations for the purchase by Ontario Hydro of the extensive assets of the Electrical Development Company, which came to be known as "the Clean-up Deal," continued throughout 1920, until a purchase agreement was reached on December 4, made retroactive to November 30. For $32,734,000 the Mackenzie syndicate sold Ontario Hydro the Electrical Development Company, the Toronto Niagara Power Company, the Toronto Electric Light Company; and the Metropolitan, Toronto and York, and Schomberg and Aurora Railways. This price, which included the bonded indebtedness and capital stock of the power companies, was several millions more than Mackenzie had been prepared to accept in 1918. There remained the Toronto Street Railway, whose franchise had expired and for which Mackenzie in 1913-14 had wanted $22,000,000. An arbitration board was appointed comprising Sir Adam Beck for Ontario Hydro, Sir Thomas White for the Company, and Hume Cronyn, MP, Chairman. The Board made an award of $13,000,000 which the city, at Beck's instigation, appealed, futilely and expensively. In October of 1924 the Judicial Committee of the Privy Council dismissed the city's appeal and allowed the Mackenzie counter-appeal, which added costs of $300,000 and interest, so that the city had to pay a total of $13,679,242.

In October 1924 another veteran Ontario Hydro employee attempted to emulate engineer Sothman and flee across the border. He was Clarence Settell, Hydro's secretary in 1907 and for nearly seventeen years the trusted confidential secretary and assistant of the Chairman. When Settell was arrested at Niagara Falls, just as he was about to cross the Honeymoon Bridge, he had in his pocket $22,000 of stolen Ontario Hydro money. An additional $8,000, part of the $29,900 he had obtained by cashing a fraudulent cheque, had already been mailed to relatives.

Settell later testified that he thought Ontario Hydro owed him that money to make up for what he had spent unofficially in entertaining influential personages in Ontario Hydro's behalf. When the police searched him they found in addition to the money a long blackmailing letter addressed to Sir Adam Beck, which was taken to the Premier, Howard Ferguson. When Sir Adam was shown the letter he declared that although it contained much obviously malicious gossip and slander, yet as Chairman of a Commission administering a public trust, he felt obliged to demand a full investigation of the charges made against himself and his administration. At a meeting of the Cabinet, Judge Colin G. Snider was appointed as a Royal Commission to make the investigation. Some thirty-nine of Settell's charges involved Sir Adam Beck personally. They accused him of conniving with irregularities, hoodwinking the other commissioners, and changing the minutes to serve his own purposes. All this Beck denied categorically. One of the witnesses before Judge Snider was F. H. McGuigan, who in 1908 had been given the contract to build the transmission line from Niagara Falls. McGuigan charged that Ontario Hydro's Chief Engineer had been forced to recommend acceptance of his bid, though it was not the lowest. After the contract was completed, McGuigan put in a bill for $412,791 more than the contract called for, and threatened "disclosures" if he was not paid. Later he sued, and got an out-of-court settlement for $86,650.

Sir Adam successfully refuted McGuigan's charges. He also defended himself and his associates successfully against Settell, whose charges involved vague allegations of graft and corruption. Judge Snider's report, rendered on December 11, was a complete vindication of Beck and of Ontario Hydro. Most of Settell's charges, said the Royal Commissioner, had been proved to be false; the others, for which some evidence existed, concerned petty administrative irregularities for which Sir Adam was not responsible and which did not involve corruption or malfeasance.

After discussing the report with his Cabinet, Premier Ferguson released it to the press, with the statement that no government action would be taken, since Judge Snider had found nothing to reflect on Sir Adam Beck or any of the Ontario Hydro Commissioners, but only on certain officials and certain aspects of Ontario Hydro's internal administration.

Sir Adam, whose precarious health could ill afford such strains, emerged unscathed from his ordeal. But his enemies, and Hydro's, were not through with him. A month after Judge Snider's report was made public, there appeared, under the eminently respectable imprint of the Smithsonian Institution, and over the signature of its Secretary, Charles D. Walcott, a publication entitled, *Niagara Falls: Its Power Possibilities and Preservation.* The seemingly innocuous title concealed a concluding section, "Ontario and the United States Electric Services Compared," which constituted in effect an unfair and unscrupulous attack upon Hydro and on Beck. Subsequent inquiry by the Federal Trade Commission of the United States revealed that Hydro's arch-enemy, the National Electric Light Association of America, had instigated the study. As Beck later declared, the Smithsonian, an institution founded a century before "for the increase and diffusion of knowledge among men," had been used for "a definitely controversial and unfounded attack upon the public activities of a great and friendly neighboring people."

The statement is contained in Beck's nineteen-page pamphlet reply to NELA, entitled, "Misstatements and Misrepresentations Derogatory to the Hydro-Electric Power Commission of Ontario. Examined and Refuted by Sir Adam Beck." Written when he was already gravely ill, it was the last public document to which he turned his hand. Tired but not tamed, Beck chose to die as he had lived, in the midst of battle for the cause he had made his own.

17

In March 1925 Hydro's stricken Master Builder submitted his indomitable will to live to the high medical court of the Johns Hopkins Hospital in Baltimore. There the doctors diagnosed his condition as pernicious anemia, a disease then regarded as incurable. They gave him six months of life.

Sir Adam, it was reported, received this verdict with his usual outraged indignation. Another enemy! Another dastardly frustration of his plans! Immediately he rushed home to build Hydro's defenses against the attacks that would certainly be unleashed when he was no longer there to repel them.

In May he saw Premier Ferguson and induced him to oppose the attempt of United States interests to develop and export the water power of the Ottawa River at Carillon. It was a foreign plot, declared Beck; another border raid. "To thus subordinate the very life blood of Canadian industry—low-cost electric power—to the dictation of foreign interests forbodes an ultimate political subserviency that no Canadian can or will tolerate."

In July, from his sickbed, he denounced Mayor Wenige of London as an enemy of public ownership and urged that a fifteen-year franchise not be given to the London Street Railway Company. The daily press reported:

> The member for London (Beck was still in the Legislature) gave evidence of his improved physical condition by announcing the policy of the Ontario Hydro Commission in regard to a new dam in the river at Springbank. He declared that the river is now only a sewer and that the approval of the scheme for the dam will not be given until the stream has been made fit to go on or in. Sir Adam Beck had a conference with Hon. Dr. Forbes Godfrey, Minister of Health, over the long-distance telephone and

157

was promised the full coöperation of the Ontario Department of Health.

To friends who gathered at his bedside Beck said: "I had hoped to live to forge a band of iron around the Hydro to prevent its destruction by the politicians."

He died on August 15. The politicians who paid him public tribute at the funeral services could scarcely wait to begin their private plotting to undermine the house that Beck had built. But they were too late. Better than he realized Beck had indeed "forged an iron ring around the Hydro." His successors inherited an organization blooded in battle, armored by a solidly welded structure of legislated powers and sanctions, and protected by the almost religious fervor of its popular support, the building of which was perhaps Beck's greatest achievement.

The defects of Beck's character contributed almost as much to his success as did his finer qualities. He was without doubt the most formidable egoist in the entire roster of Canada's great men: opinionated, intolerant, and violently irascible. Ruthless to his enemies, he treated even his best friends as expendable, when necessary to further public ownership's greater good. His judgment was sometimes at fault, as in the long and costly radials controversy, a losing battle which he was still fighting on his deathbed. But since he never doubted either himself or his cause he was able to recruit an army of followers whose fanaticism often matched his own.

Fortunately, Beck was genuinely disinterested. He wanted power not for himself, but for Hydro. As a dictator he was never perverse or irresponsible. It was merely that he lacked any measure of humility or self-doubt. From his associates on the Commission and from Ontario Hydro's employees he demanded no more than he required of himself; merely inordinate industry, unqualified loyalty to Hydro, and complete agreement with Ontario Hydro's Chairman, whether he happened to be right or wrong.

Fortunately, Beck was right most of the time. He was right in his vision of the Province's industrial growth and in the essential role in its development played by Hydro's power-at-cost. He was right, since co-existence was obviously not possible, in fighting the private utilities —although it would have been wiser and less costly if he had tempered some of his victories with mercy. He was right about buying and developing hydro-electric horsepower wherever it could be had, rather than be forced to accept the costly alternative of steam power. He was right in backing the young engineers who, with comparatively little prior experience, drew the plans for what was then the largest hydro-electric development in the world, at Queenston-Chippawa. A smaller man would scarcely have dared so big a gamble.

Beck was above all right in keeping political placement out of the Ontario Hydro organization and refusing, as he did, to make appointments and promotions based on any kind of influence or political preferment. At the same time, of course, he kept Hydro *in* politics by encouraging the organization of pressure groups formed to elect political friends and punish political enemies. But to call this reprehensible is to reject the democratic process. Some of Beck's pressure groups, like those organized to promote the radials, helped to implement Beck's own wrong-headedness. Others, like OMEA, were necessary to protect Hydro from its enemies and to support its program during the critical early years until it had a chance to justify itself by its works.

For over twenty years, from 1903, when Beck became a member of the Ontario Power Commission with E. W. B. Snider as Chairman, to within a few weeks of his death, Beck was Hydro and Hydro was Beck. It was Beck who drafted the Act of May 14, 1906, which created Hydro, and whose hand was apparent in most of the subsequent legislation affecting the Commission. It was Beck whose speeches rallied the voters to pass the by-laws that brought the municipalities into the Hydro partnership. And it was Beck who threw back the periodic offensives of the private-power lobby with documented defense and savage counter-attack.

Beck's attitude toward the heads of the provincial governments, although he sometimes lacked cabinet rank, was invariably that of an independent potentate whose powers and perquisities must be respected and whose drive to expand must be supported without question. None of Beck's successors dared to behave with anything approaching his arrogance. But then, none of them were confronted with Beck's problems, and none quite matched his achievements. Beck's best friends frequently found his manners intolerable, but even his worst enemies acknowledged his personal integrity and his public dedication.

During World War I Beck's German ancestry had made him the target of many vicious attacks. He proved his patriotism by working overtime for the army without pay as Remount Officer; this without delegating or neglecting his responsibilities as Ontario Hydro's Chairman during the most critical period of its expansion. During these years Ontario Hydro assumed control of the Electric Power Company, constructed its first generating stations, greatly extended its transmission lines, projected the Queenston-Chippawa development and tripled its power load to meet the insatiable demands of Canada's war industries. Undoubtedly Beck's prodigious wartime activities, which left him chronically exhausted, served to hasten his death.

To Beck, as much as anyone else, Canada owes the St. Lawrence Power Development and Seaway, which he projected in 1913 and advocated in season and out with a stubborn faith not shaken by repeated frustration and disappointment. Complete utilization of this large source of power was, he knew, ultimately essential if Hydro, in years to come, was to continue to provide all the electricity needed by Ontario's industrial and domestic consumers. To one of his employees he confided what was seemingly his chief concern when he knew that he had not long to live: "Remember what I am telling you," he said, "They have no cause to raise Hydro's rates. Watch what they do when I am gone."

As Beck was well aware, "they," meaning the politicians whom Beck detested as a class, although he was certainly one of them, would be plotting this and other mischief, even as they composed the eulogies that they would speak at his funeral. To many of the silk-hatted dignitaries who stood at the grave of Beck dead, Beck living might well have said, as he said to one of those who visited his sick room: "What in hell are you doing here?"

What brought them, of course, was the fact that Beck, undoubtedly one of history's great curmudgeons, was also one of Ontario's greatest public benefactors and one of Canada's greatest statesmen. Among the more discerning of the obituary tributes paid to Beck was that of his friend and biographer, W. R. Plewman, which appears in his book, *Adam Beck and Ontario Hydro*:

Adam Beck was a great man who would have achieved greatness in any country in which he might have spent his life. Judged solely from the standpoint of the material benefits that he brought to his countrymen, he may deserve to rank as Canada's greatest son. No other public man prominent in the history of the Dominion possessed a combination of such dauntless courage, such extraordinary far-sightedness and shrewdness, such organizing ability and such irresistible resolution. No other man was surrounded by so many powerful and unscrupulous opponents. Other men matched him in the loftiness of their patriotism but few if any equalled him in sacrifices made for the sake of the common people and for sufferers from tuberculosis. He was a hard man and sometimes brutal. He was anything but pleasant in a number of his personal contacts. But let this be said to his everlasting credit that a man of greater refinement and tenderness could not have mastered the alliance between predatory interests and pliant politicians and given Ontario the cheapest hydro-electric power in the world and the greatest publicly owned power system.

To his successors and to the people of Ontario, Hydro's Master Builder bequeathed a living creation: a network of generating stations and transmission lines that already was beginning to nourish and fructify the farthest corners of the province with the life blood of the people's power; an organization shaped, hardened, and tested by the hand of the master, powered in perpetuity by the dynamo of a great tradition.

More than anything hewn in stone or cast in bronze, Hydro will remain Beck's living monument.

18

Even before Hydro was born, and increasingly during the first twenty years of its existence, Adam Beck had fought the private-power interests: in print and on the platform; in Canada, in London, and in the United States. In 1925, as he lay dying, the newspapers announced the publication of *Niagara in Politics* by Professor James Mavor, the retired head of the Department of Political Economy at the University of Toronto and a scholar of world renown for his economic works on Russia and Western Canada.

Almost a decade earlier, in 1916, Mavor had written a series of articles attacking Hydro for the *Financial Post* in Toronto, one of Hydro's most persistent critics. In *Niagara in Politics*, Mavor brought his appraisal up to date and concluded that Hydro, far from being a genuine experiment in public ownership, was actually "an attempt on the part of a small number of politicians to establish an industrial monopoly and to manage this monopoly in such a way as to keep themselves in power. In order to effect this object they have violated constitutional law and practice; they have assumed absolute authority; they have closed the courts of justice against proceedings adverse to themselves, and they have encroached upon the liberties of the people."

There is evidence that Mavor was moved not only by his conviction that public ownership of utilities was economically unsound and politically iniquitous but by a deep personal dislike for Sir Adam Beck. Whatever the circumstances, the academic reputation of the author and the affection with which he was held in Toronto made *Niagara in Politics* a highly effective propaganda weapon in the hands of Hydro's enemies.

If Beck could have but known it, that was the last big gun of the propaganda barrage which, with few intermissions, had beset Hydro almost from the moment of its inception. The fight was by no means over, but in Canada Hydro had weathered the storm, while across the border Hydro's historic enemy, NELA, was itself coming under increasing fire.

In 1929, the panic year that saw the collapse of the paper empires built by the public-utility holding company promoters. *Electrical Utilities, the Crisis in Public Control*, by William E. Mosher and others, made its appearance. It was issued under the auspices of the School of Citizenship and Public Affairs at Syracuse University without benefit of private-utility subsidies. Professor Mosher and his associates had taken the trouble to get the facts straight about Hydro and its rates, and about the private power companies in a roughly comparable area of New York State. They concluded:

> The Ontario municipalities are on the whole enjoying more favorable rates for electrical power than selected cities in New York State which are served by private companies. . . . A further conclusion is that there seems to be no sound basis for the repeated contention that large industrial consumers are penalized in Ontario by subsidizing small users, at least to such an extent that the practice brings their costs above the average that is paid in New York cities. . . . Insofar as comparative costs are an index to the managerial efficiency of an enterprise, the tables submitted . . . go to show that the Ontario Hydro Commission has nothing to fear from this test. . . .
>
> The Ontario system has . . . demonstrated the feasibility of public ownership and operation of this utility on a thoroughly businesslike basis and on a statewide scale. It demonstrates that it is entirely possible for the government to secure the services of competent managers and technicians if it is willing to pay adequate salaries and make appointments on merit. . . . It is far from axiomatic that government enterprise is necessarily and inevitably inefficient. . . . It is not inconceivable that the Ontario experiment may serve as a beacon light to the people of the United States, should a wave of protest ever get under way at the methods of those in control of the industry on this side the Canadian border.

In 1931 Ernest Gruening, an editor of the *Nation*, who was to become Governor of Alaska, published *The Public Pays, a Study in Power Propaganda*. The book abstracted the testimony and findings yielded by a three-year study undertaken by the U.S. Federal Trade Commission. The FTC study supplemented an investigation by a Committee of the United States Senate, conducted as a result of a speech delivered by Senator Thomas J. Walsh of Montana on March 28, 1927.

Senator Walsh urged the Senate to appoint a five-man committee to investigate the capitalization of the power and light industry, its control by holding companies, and the extent to which the private-power interests had attempted to influence public opinion against public ownership of power utilities.

Despite powerful opposition the committee was appointed. Over a period of three years it accumulated 14,000 pages of testimony. Its report was a devastating exposure of the efforts of the electrical utility industry to control public opinion, make friends, and suppress and punish enemies. This "public relations" campaign of the industry was launched in 1919 by Samuel Insull, whose holding-company empire collapsed during the depression and who spent his last years in exile. Within two years, Insull's creation, the Illinois Committee on Public Utility Information, distributed five million pieces of pamphlet propaganda to newspaper editors, bankers, businessmen, teachers, preachers, librarians, politicians and public officials of all sorts. The Committee conducted a Bureau which supplied Chambers of Commerce, clubs and churches with speakers holding "sound" views on the public ownership of electric utilities. It provided high schools with free "educational" literature on the subject.

The Illinois Committee was the progenitor of the National Electric Light Association (NELA) which conducted similar activities on a larger scale and coördinated the efforts of the state propaganda committees. M. H. Aylesworth, NELA's managing director, suggested at a utilities convention that occasionally it would be useful to "retain" a college professor at the bargain rate of $100 or $200 a year for the privilege of "studying and consulting with him." "Don't be afraid of the expense," urged Aylesworth, "the public pays." John B. Sheridan, Secretary of the Missouri Committee on Public Utility Information, echoed this sentiment, but pointed out that there was also the problem of textbooks. "Teachers come and go. Textbooks remain." Sheridan became chairman of NELA's subcommittee on textbooks and author of an electrical *index librorum prohibitorum*. Some textbooks, he subsequently reported, had to be suppressed by quiet diplomatic measures. Others were revised by the authors, who were usually amenable to suggestion since, as Sheridan observed, "This gives the boys a chance to write new textbooks and make some more money."

One of the many "retained" teachers was Professor E. A. Stewart of the University of Minnesota, who conducted an allegedly impartial, scientific and scholarly survey of rural electrification in Ontario. His conclusions, syndicated by a chain of newspapers that used NELA propaganda, were that the farmers in Ontario were very much worse off than those served by the privately owned utilities. Ontario Hydro's engineering department, he declared, had checked the facts on which he based this conclusion.

To this, Ontario Hydro Chairman Magrath replied that Stewart had left Toronto before the check was completed; that changes and additions requested by the Commission's engineers were never made; that many of the figures used and the statements made in Stewart's report were not in accordance with the facts. Stewart, who had been on the payroll of the private-power corporations while making his study, soon resigned from the University of Minnesota to become President of the Minnesota Power Company.

In addition to propagandizing through all the mass media: newspapers, magazines, films, radio, music publishers (some 75 utility-plugging songs were composed and distributed), NELA and its member committees spent much time and money in hounding and suppressing the advocates of public ownership. One of the victims of this organized vendetta was Carl D. Thompson, Secretary of the Public Ownership League of America, who in a speech at Charles City, Iowa, had compared Hydro's rates favorably with those charged by privately owned corporations.

The silencing of Thompson became a major NELA objective. All across the country he was harassed by swarms of private utility hirelings. His meetings were boycotted. His addresses were cancelled and "entertainers" provided to replace them. Even that was not enough, thought John B. Sheridan of the Missouri Public Relations Committee. The best way to have treated Thompson, Sheridan said, "would have been to throw him in a ditch and to hell with him."

Sheridan, the most zealous of NELA's propagandists, did not hesitate to crack down on R. R. Parks of the College of Agriculture at the University of Missouri. "Step softly on the electrification of farms in a wholesale way," he warned, "unless you have the approval of the local power companies affected."

Later he suffered a crisis of conscience. "Yes, men are a little breed," he wrote the editor of a public-utility magazine. "Possession of property breeds liars and cowards. The man who invented private ownership was a mortal enemy of the human race. For thirty years I spoke as I felt. For five years I held my tongue. Now I mean to resume the greatest of human rights—that of free speech. Damn it all, John, they can never make hypocrites and cowards of all people. . . ."

Three years later the unhappy man committed suicide. But NELA had quickly replaced him with a less sensitive mercenary, and its public relations campaign continued with redoubled force. In its issue of

December 4, 1927, the *Boston Herald* published a full-page Sunday feature entitled *Here is the Truth about Ontario Hydro Electric*. The article presented the kind of "truth" that was designed to serve NELA. Professor Mavor and Carl D. Thompson were quoted in parallel columns, but with a two-to-one edge given to Professor Mavor, whose anti-public-ownership views were further reinforced by plentiful quotations from Professor E. A. Stewart and other paid NELA propagandists. The article cost NELA $1,000, for which the *Herald* provided 10,000 reprints of the issue at ten cents a copy.

Of those quoted in the *Herald* article, all but Thompson of the Public Ownership League were subsidized. F. G. Gordon, who edited the copy, was a paid NELA propagandist. Professor Mavor, whether or not he was cognizant of the source, had received money from the Association. Dr. E. A. Stewart of Minnesota was on the payroll. And Dr. A. E. Kennelly, Professor of Electrical Engineering at Harvard University and the Massachusetts Institute of Technology, on whose opinions Gordon's article was partly based, had been retained by NELA as a technical adviser.

Commenting on this and other examples of NELA's propaganda methods, Gruening writes in his concluding chapter:

> Whether the industry has the wisdom and the foresight to reform itself from within, to set its own pirates adrift . . . to cease its misleading propaganda, to rise to its full responsibilities and duties as a *public* utility and honestly seek to achieve the objective of service at cost, a fair return—and no more—on capital honestly invested, remains to be seen.

The industry did reform itself in a measure, largely because it was forced to do so by Congressional investigations and Federal prosecutions; also by the necessity, after 1933, of competing with the power yardstick set up by the Tennessee Valley Authority, fathered by Senator William Norris and endorsed by President Franklin D. Roosevelt. While he was Governor of New York State, the President had become a close student of Ontario Hydro and many of its features were incorporated in the American legislation which brought the Authority into being. TVA, for example, used Hydro's power-at-cost formula to electrify farms, process phosphate ore at Muscle Shoals, stimulate local industry and otherwise rehabilitate the impoverished physical and social landscape of the Tennessee Valley. Like Hydro, TVA was obliged to fight the private-power lobby with one hand, while restoring with the other the fertility of the Valley's soil, and the health and hope of its people. Like Hydro, TVA applied successfully the principle of "power at cost, power to the remotest hamlet." Like Hydro, TVA won worldwide acclaim as an example of democracy at work.

One of TVA's original directors, who was for a time Chairman of its Board, was the Canadian-born Harcourt A. Morgan, a graduate of the Ontario Agricultural College at Guelph. During the nineties Dr. Morgan served as state entomologist for the State of Texas where his pioneering entomological studies led to the conquest of the Mexican cattle tick. Prior to his appointment by President Roosevelt as a TVA director he was President of the University of Tennessee. As director of TVA's agricultural program his evangelical zeal in bringing the benefits of electrification to the impoverished farmers of the Tennessee Valley rivalled that of his great precursor, Adam Beck.

19

When Sir Adam Beck died, Ontario Hydro was the largest publicly owned hydro-electric utility in the world—and it had just begun to grow. To his successor Hydro's master builder bequeathed an organization forged in the heat of public controversy and tempered by twenty years of trail-blazing development and expansion; nevertheless, the problems that confronted Ontario Hydro's second Chairman were no less arduous and difficult than those which had faced his predecessor. Under Beck's belligerent, crusading leadership, Hydro's enemies had been repulsed and its daring foresight justified, but there remained to be solved problems of management and organization, and new supplies of power to be obtained, as Ontario, in common with other parts of North America, entered on the most dynamic period of industrial expansion the world had yet experienced.

Charles A. Magrath, Chairman of Ontario Hydro from 1925 to 1931, was as restrained and self-effacing in his public behavior as his predecessor had been flamboyant; the only similarity between the two men, in fact, lay in their dedication to the cause of public ownership. Born in the village of North Augusta in Eastern Ontario in 1860, Magrath went west as a youth of eighteen to become one of the builders of Western Canada. A surveyor and administrator by training, he became an authority on irrigation and hydraulic resources, and acquired a national reputation in those fields. Entering politics in 1891, he served first as mayor of Lethbridge, later representing the city in the Territorial Assembly of the North West Territories. From 1908 to 1911 he sat in the Dominion House of Commons representing Medicine Hat, after which he was appointed as one of the first members of the International Joint Commission, and from 1914 to 1935 he was Chairman of the Commission's Canadian Section. A member of the War Trade Board, he was appointed Dominion Fuel Controller in 1917. He also served on many other boards and public bodies, among

168

"Betwixt the Lakes of Ontario and Erie, there is a vast and prodigious cadence of water which falls down after a surprising and astonishing manner, so much so that the Universe does not afford its parallel."

So wrote Father Louis Hennepin, Recollet missionary-explorer-writer, member of LaSalle's first expedition, on December 6, 1678. The "prodigious cadence" and famed scenic marvel was to inspire a dream—a dream of harnessing the descending waters to provide power for industry.

In the beginning it was a story
of repeated efforts to tap
the tremendous potential of
the thundering waters.

Opposite, the first diversion of
water for power purposes on the
Canadian side was used by a
generating plant of the
International Railway Company
in 1893.

Below, near St. Catharines,
the original DeCew Falls GS
began operating in 1898 and was
acquired by Hydro in 1930.
The new plant was completed
in 1947.

As solutions were found to the
technical problems involved,
a new issue rose: power at cost,
publicly owned power.

Below
P. W. Ellis was first Chairman of
the Toronto Electric Commission,
formed in 1903.

Opposite
D. B. Detweiler of Berlin and
E. W. B. Snider of St. Jacobs
both took up the cudgels in
behalf of public ownership.
Detweiler cycled the country
roads winning converts.

On this page, steel towers and
cables binding the municipalities
together.

The Ontario Power Company plant at
Niagara Falls (*opposite*, shown
as it appeared in 1909) was the
first supplier of power to the
14 Southern Ontario municipalities which originally signed
agreements with Hydro.

This group photographed near
St. Thomas in 1913 is a forerunner of present-day survey crews.

The press took sides and hot words were used for
and against publicly owned power. A contemporary
the temper of the times. Hydro's first contracts with the
participating municipalities were signed May 4, 1908.

A SPRINGTIME IDYLL.

MR. W.—IF THEY TELL YOU, BECKY DEAR, THAT I AM COLD-HEARTED TOWARD YOU, THAT I
AM PERFIDIOUS AND UNTRUTHFUL, DON'T YOU BELIEVE IT! I AM WHOLLY THINE, M
BECKY, WHOLLY THINE!!

Detweiler's home town, Berlin (now Kitchener)
conducted its Switching-On ceremony in a big way.
Leading the motor cavalcade into the town were
Mayor C. C. Hahn, Dr. H. G. Luckner (MPP for
Waterloo North), and Ontario's Prime Minister
Sir James Pliny Whitney.

Bottom of opposite page
Left to right are Hon. A. J. Matheson, Provincial
Treasurer; Mrs. Beck and Hon. Adam Beck, first
Commission Chairman; and, seated beside the driver,
C. H. Mills, later MPP for Waterloo North.
Berlin's streets were festooned with electric
lights in readiness and crowds lined the streets
waiting for the moment of switching on.

Miss Hilda Rumpel carried a velvet cushion on which reposed the push-button for the ceremony (now in the Waterloo Historical Society Museum). Standing by in Hydro's Berlin Transformer Station, Charles Sheppard awaited the signal. Press headlines emphasized the emotional moment when the people's power became a reality.

What's the News? Read the RECORD

THE BERLIN

THIRTY-THIRD YEAR.   BERLIN, O

WHITNEY AND B

Sir James Whitney Declines To Deprive Hon. Adam Bec
Both Perform the Important Function, the Premier Gro
an Act of Courtesy Which Brought an Outburst
Falls---Formal Turning on of Niagara Power
of About 8000 People, Including Disting

The triumphal achievement of the greatest public undertaking in the history of the Dominion of Canada was signalised in Berlin yesterday, when, by pressing a button, Premier Sir James Whitney formally turned on Niagara Power in Ontario. True it is that the power has been in use in Berlin for some weeks past, but not in an official sense.

That Berlin should be chosen for this signal honor is but fit and proper. It was here that the inception of the Niagara Power movement took place, and it was here that probably the greatest interest has been taken in the movement. Naturally when a public ceremony was suggested Berlin put forth its claim for the honor, and those claims were of such an emphatic nature that the Government and the Commission rightfully decided that this town was the place in which any formal ceremony should be held. Financially assisted

filled coaches pulled into the station. Here the reception committee again welcomed the prominent men, which included the Premier and several members of his Cabinet.

The Procession.

A procession, was then formed, headed by the 29th Regiment Band, followed by a long string of automobiles; the Galt Kilties, another string of autos, the Preston Band, and still more autos. The procession was an imposing one, and in the carriages were men who have worked out the power scheme to a successful issue, and also the men who are to-day directing the affairs of the Provincial government. As stated above, never before in the history of the town has there been a greater gathering of such distinguished visitors.

On arrival at the market square the bands dropped out of the procession and

The Banquet in the Evening Was th
Probably in Ontario--Sir James
Members of Parliament, Ma
and Hundreds of Other
Completion of a Gi
Day in

Waterloo,--to be exact the town of Berlin has the honor of being the birth place of the distinguished, persevering and ultimately successful Chairman of the Hydro-Electric Commission, the Hon. Adam Beck. If for no other reason the people of this County will rise on this occasion to do him the honor which he least a pc Province joy ment o er for ne purposes. by thems achieveme promoters proud, an

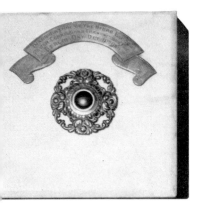

TOBER 12   NUMBER 239

PRESS BUTTON

lege of Formally Turning On Niagara Power, and

ck's Hand and Pressing His Finger on the Button,

llmost as Voluminous as the Roar of Niagara

as Successfully Carried Out in the presence

from the United States and Canada

Ever Held in Berlin, and
Beck, Cabinet Ministers,
n, Newspaper Men,
e the Successful
he Greatest

18

le foresaw the great influence the
resources of the falling waters of
the province would have upon the e-
nomic development of the province
And while the "government may first
consider the water power resources
of the province as a source to the
while treasury, the furnishing of
hydro-electric power to the people was
ost was paramount in his mind, and

our work to supply these demands.

What Winnipeg has Done.

"When we think of Winnipeg, a
small city compared with Toronto,
spending three and a half millions
for power, shall we in Ontario doubt
that we shall ultimately need all the
power we can secure," said Mr. Beck
(Cheers.) Toronto's share of the
cost of the hydro-electric line would
be $850,000, as compared with an
expenditure of over three millions,
which the privately owned line had
cost into Toronto.

"Therefore," he said, "we have no
fear of competition."

Mr. Beck said the work of the commission had only begun. "We shall
not rest," he said, "until we have no
more coal oil, no more gas, and I
hope, no more coal. (Cheers.) Mr.
Beck thanked the municipalities for
their confidence and co-operation.

"Sir James Whitney," Mr. Beck
continued "is present to perform

truthfully say that we have never sold
the English investor bad securities.
We have given him good value for his
money, and we must continue to do
so."

"Mr. William Mackenzie, on July
16th, 1909, having just returned from
England, said:

"There is, as far as I could discover,
no sentiment against Canadian secur-
ies in London as the result of any legislative action which has been taken in
this country. If there is, I did not meet
those who could have told me about it.
It is all a matter of price and our securities bring what they are worth in
competition with other offerings."

"Mr. E. B. Osler, on Nov. 29th, 1909,
said:

Continued on page 2.

SEVERAL STARCHERS WANTED
AT ONCE.

Steady employment all year round.

In Toronto in 1911
a miniature Niagara was
erected over the main
entrance to the City Hall,
but this impressive display
backfired. At the moment
when the switch was pressed,
the electricity went on
according to plan but water
drenched the assembled
dignitaries. London and
Port Arthur also held
important switching-on
ceremonies.

Systems were enlarged;
engineers were called upon
to design new distribution
stations and feeder lines;
more and more men were
employed as the demand for
power multiplied.

Now there was a new emphasis; after the political
battles, after the technical problems, now
the stress was on work, the actual labor of
construction, tracing the skeleton of the system,
extending the lines to take power across the miles.

Above, at the construction of Eugenia Falls GS in
1914, steam shovels were operated by steel chains,
"continental" dump cars were hauled by "dinkey" engines.

Opposite, inspecting construction on the
Queenston-Chippawa GS with Sir Adam Beck are
Dr. F. A. Gaby, Chief Engineer (*left*),
and H. G. Acres, Hydraulic Engineer (*right*).

Queenston-Chippawa GS was the largest hydro-electric
plant in the world at its completion; in 1921, the official
opening was attended by Sir Adam's daughter
Marion Beck and Ontario Premier E. C. Drury. Later,
it became known as Sir Adam Beck-Niagara GS No. 1.

"If I have helped to lessen the household cares of
the housewife by making electricity her servant . . .
if I have helped the farmer to make life more attractive,
to help keep the boys and girls on the farm, then I have
not labored nor have you coöperated with me in vain."
Above, the farm of M. W. Keefer, near Galt, was among the first
to receive electricity in Ontario.

Opposite, an electrical milking demonstration at the farm
of Mr. and Mrs. Alex Anderson in St. Thomas was
attended by Hon. Adam Beck (second from right).
Beck's drive for "power to the remotest hamlet"
used a travelling Hydro "circus" to demonstrate
the benefits of electrical farming.

Opposite
Cables linked this motor wagon to
the Hydro transformer wagon for
a corn-cutting and silo-filling
demonstration in Oxford County
in 1912. Today electricity is
available to 95 per cent of
Ontario's farms–a record of
rural service unsurpassed anywhere.

Ontario's expansion as an industrial province has closely paralleled the development and growth of Hydro.

Left, power serves the pulp and paper industry.

Opposite, above, molten gold is poured from an electric furnace.

Below, a modern electric furnace at the North American Cyanamid Limited plant at Niagara Falls pours molten calcium carbide.

Opposite
A large electric furnace
is tapped in an
Ontario steel mill.

This page
A suspension block or
"traveller" facilitates
the construction of
transmission lines.

Hydro is an important employer of labor throughout the province, and as the demand for power stimulates expansion, low-cost power stimulates the development of industry.

Below, Abitibi Canyon GS.

On this page is pictured Caribou Falls GS.

Opposite, bottom picture shows
Pine Portage GS; *top,* Otto Holden GS
was formerly La Cave, Ottawa River. Other
stations renamed in 1950 in tribute
to the men who made outstanding
contributions to Hydro's success included
Queenston-Chippawa, which became
Sir Adam Beck-Niagara GS No. 1;
Westminster, renamed E. V. Buchanan TS;
Peterborough, which became Ross L. Dobbin
TS; Windsor, now J. Clark Keith GS;
Sudbury, renamed R. H. Martindale TS.

Opposite
Top, George W. Rayner GS,
formerly Tunnel GS,
was also renamed in honor of
one of Hydro's builders.

Below,
Stewartville GS is shown.

Richard L. Hearn GS at Toronto is
Canada's largest thermal generating station.

Below, A. W. Manby TS and Service Centre
comprises central stores, central garage,
and other services.

With the waters of the
province exploited almost
to the full, Hydro is turning to
thermal and nuclear development.

This page shows models of the
Nuclear Power Demonstration
near Chalk River, and the
Douglas Point Nuclear
Power Station.

Operating on a 25-cycle frequency
in the beginning, Ontario Hydro faced
a tremendous problem in converting
all electrical appliances in Southern Ontario to
60-cycle power. Frequency standardization
was carried through the 1950's.
In July,1959, on completion of the program,
over seven million appliances had been
converted for more than a million customers.

Hydraulic models of generating stations have saved much time and money, and permitted valuable experimentation. This one relates to the redevelopment of Niagara power made possible in 1950 through the Niagara Diversion Treaty, which provided for maximum power development while preserving Niagara's scenic beauty.

This model was used by both
Ontario Hydro and New York's
Power Authority, partners in
the long-delayed development
of the St. Lawrence River
for hydro-electric power.

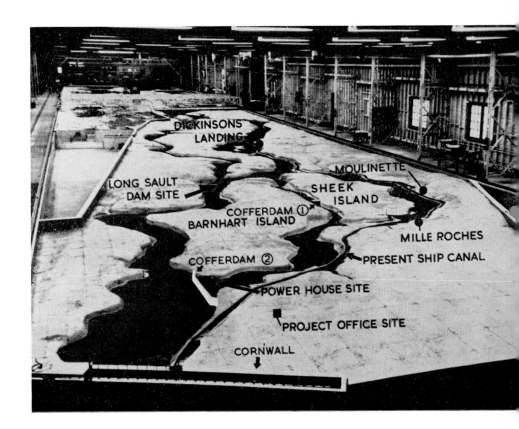

Construction of the
Sir Adam Beck-Niagara GS No. 2
began within weeks after the
Niagara Diversion Treaty was
ratified on October 10, 1951.
Preliminary earth-moving
is shown on the Niagara
River Gorge, upstream from
Sir Adam Beck-Niagara GS No. 1.

Concrete was pumped through
steel pipes for this 45-foot
diameter tunnel beneath
Niagara Falls, Ontario.

Opposite
Aerial views show the
powerhouse area of
Sir Adam Beck-Niagara
GS No. 1 and No. 2,
open cut canals and
crossover.

Left
Close-up shows trench
excavation for the
penstocks, through which
water drops down some
295 feet to the turbines.

Twin five-and-a-half-mile tunnels of
the Sir Adam Beck-Niagara GS No. 2
surface at the trapezoidal section
of the open-cut canal leading to
the headworks of the powerhouse.

Aerial views show the crossover of the two canals, and the
Pumping Generating Station
near Queenston, which has a
capacity of 170,000 kw.

Construction is underway
at the Niagara River
Control Dam, part of
the Niagara Remedial
Works program for
preserving the beauty
of the Falls and
effecting the most
efficient use of
Niagara waters.

Completed in 1958, Sir Adam Beck-Niagara
GS No. 2 has sixteen 75,000-kw units.
With its associated pumped-storage plant
it has a capacity of 1,370,000-kw.

The St. Lawrence Power Project at last.

Opposite
Two steel tunnels were built to allow ships to pass above,
with men and building materials going through below.
The larger tunnel could take truckloads up to 15 tons.

Ground-breaking ceremonies in August 1954; first sods
were turned by Canada's Prime Minister Louis St. Laurent,
Thomas E. Dewey, former Governor of New York,
and Ontario's Premier Leslie M. Frost.

Two cofferdams were
built to allow
construction to
proceed "in the dry."

Excavating for the powerhouses and concreting for Robert H. Saunders-St. Lawrence GS shown in progress. By March, concreting was half completed.

Opposite,
Two views of the
Long Sault Dam,
a key part of
the St. Lawrence Power Project,
and (*bottom picture, right*)
Iroquois Dam are shown.

This page, houses were moved
to new locations, and new
roads were constructed,
new towns came into being.

New schools had to be built and
new churches rose to take the place
of the old. *Below*, close-up of
installation of a 225-ton rotor.

Opposite, above, the cofferdam had been
breached and partly removed in May 1958.

Below, before dynamiting the cofferdam.

Thirty tons of explosives released the waters of the
St. Lawrence into the headpond area, upstream from
the International Powerhouses of the St. Lawrence Power
Project. The two powerhouses—one built by Ontario
Hydro, the other by the New York State Power Authority
—merge at the International Boundary to form one
continuous structure 3,300 feet long.

An international four-way handshake symbolizes the joint efforts that planned and executed one of the epic engineering achievements of the twentieth century. In June 1959, Queen Elizabeth visited the plant to unveil a tablet to international friendship.

Men have dreamed, men have planned, men have toiled to bring to life the vision of those who first saw the potential of Hennepin's "prodigious cadence." Lovely Lake St. Lawrence provides a new playground, created by the flooding of the headpond on July 1, 1958. Ontario Hydro played a dramatic part in easing the burden of labor in industry and on the farm, and in modernizing the homes of a vast province. That Adam Beck–and Ontario Hydro–did not toil in vain, is attested through the length and breadth of Canada's richest province.

them the Ontario Highways Commission, 1913; the committee to investigate agricultural conditions in Alberta, 1920; and the Royal Commission on Newfoundland, 1933. In 1940 he earned the highly prized Julian C. Smith Medal of the Engineering Institute of Canada, of which he had been a member since 1917.

Of Beck's colleagues on the Commission appointed during the Drury administration, Colonel Carmichael resigned when G. Howard Ferguson became Premier in 1923, and J. George Ramsden, an anti-Beck Liberal appointed to succeed the recently deceased Fred C. Miller, was removed from the Commission by order-in-council on his refusal to resign. He was succeeded by the Hon. J. R. Cooke, who in 1920 had sponsored the flat-rate proposal for rural power districts. Since no third Commissioner was appointed during Beck's long illness, it frequently fell to Cooke to assume full responsibility during 1924 and 1925. Upon Magrath's appointment to the chairmanship, C. Alfred Maguire, a former mayor of Toronto and staunch Hydro partisan, was named as the third Commission member to represent the OMEA. Fred Gaby continued as chief engineer and Major Pope as secretary.

From the outset of his administration, Magrath made it abundantly clear that there would be no break in Hydro policy. On the contrary, his early reports to the Legislature contain many passages, such as the following, which might have been written by Beck himself:

> The various electrical systems are being operated for the benefit of the people, with the Commission as the central coördinating trustee acting for the municipalities who have combined to work their electrical properties in coöperation . . . I would recall that the Commission's business is not only to supply power at cost, but at as low a cost as is consistent with the maintenance of a highly efficient equipment and the provision of proper safeguards in the way of reserves. . . .

Nevertheless, subtle differences can be sensed in Ontario Hydro's corporate personality from this time on. Gone is something of the lusty exuberance of earlier days and along with it much of the excitement it engendered. During the last few years of Beck's administration Hydro had in fact become big business and as such called for the management techniques which Magrath's character and experience enabled him to bring to it. Just how big Hydro had become by 1926 was set forth in an article contributed by the new Chairman to the Toronto *Globe* toward the close of the same year. It appeared in the paper's annual Financial Review and said in part:

In 1925, the total capital investment of the Hydro-Electric Power Commission in the undertaking under its immediate jurisdiction was approximately $199,000,000. In 1926, the corresponding total is $204,000,000, showing an increase of about $5,000,000. The total investment of the municipalities in connection with their local Hydro utilities now aggregates about $81,000,000 so that the total investment of the Commission and the municipalities in the whole undertaking is about $285,000,000.

The combined reserves and surpluses of the Commission and municipalities total about $55,000,000. In this connection it is interesting to observe the substantial amount of these reserves compared with the total capital expenditure and, further, to note the fact that the larger proportion of this amount has been accumulated during the last few years. . . . From now on the reserves and surpluses will accumulate at a strikingly rapid rate. In fact, this amount will approximate $8,000,000 to $10,000,-000 per annum.

This is the provision in the Commission's program through which the municipalities eventually will possess a hydro-electrical undertaking which has had its capital outlay completely refunded and this out of the relatively low rates paid by its customers—rates which for similar service over such a territory are not paralleled elsewhere in the world. . . .

Along with financial giantism, the physical growth of the Hydro system had been equally impressive. Of Ontario's 3,000,000 population, 2,200,000 were now supplied with electricity by the Commission and its 249 associated municipalities through 400,000 individual customers, and nearly 2,300 miles of rural lines had been constructed or approved to bring the boon of light and labor-saving devices to some 19,000 customers located on farms or in hamlets. In addition to optimum production in the Niagara System, massive engineering works had been undertaken in the Thunder Bay district to develop hydro-electric sites there and to provide storage dams for the year-round maintenance of stream flow, and in the Georgian Bay district the Hanna Chute plant on the Muskoka River was placed in operation, part of a project to exploit to the utmost the hydraulic resources of a relatively restricted area.

In the contemporary framework of the mid-1920's, Hydro's success was rightly regarded as phenomenal from every conceivable point of view—political, financial, economic and social. A daring experiment had been brought into being by the people of Ontario through democratic action, and had proved that such an enterprise was not only possible but in the public interest. There had been minor scandals, it is true, and politics had played a part in some appointments, but both the honesty and efficiency of the management had been above reproach. Whether by appointment or employment, the men connected with Hydro had served with great ability and integrity, and the people of Ontario could point with pride to their creation of an enterprise that had become an object of intense interest the world over.

Despite these pinnacles of achievement, those attained during the five years of Magrath's chairmanship of the Commission were to reach much higher levels, particularly with respect to the financial structure, management, and the provision and distribution of electricity. The explanation is contained in the post-war boom of World War I. Across the border at Detroit, the American automobile industry produced 3,602,540 motor vehicles of all types in 1924; 4,300,934 in 1926, and 5,337,087 in 1929. These figures reflect the economic effects of the two great nineteenth-century technological advances: the invention of the internal combustion engine in mid-century and, in its closing decades, the generation and long-distance transmission of electricity. Partially exploited by venture capital before 1914, these twin production tools, under the exigencies of World War I, were subjected to costly research, development, and forced expansion which otherwise might have been delayed a generation. The effects, by the mid-1920's, were seen in a second industrial revolution whose consequences have been much more far-reaching than the first, when the steam engine replaced animal and water power. In response to the dynamic stimuli, new demands had been established, increased purchasing power created and industrial production throughout the United States and Canada had spiralled upward.

But in Ontario, and generally throughout that vast area of Canada embraced by the Pre-Cambrian Shield, a phenomenon had taken place for which no exact parallel can be found elsewhere. Thanks to the internal combustion engine and its proliferations, the Shield had lost both its remoteness and inaccessibility. Now there were surplus wartime airplanes that could be bought for a song, and a new breed of men to fly them, the Canadian bush pilot. And when a likely mineral prospect was located beyond the sixtieth parallel or in the region of Hudson Bay, it was no longer necessary to tote in supplies and equipment laboriously by canoe and tump line. There were self-propelled caterpillar tractors, the lineal descendants of the first clumsy Somme tanks, which could haul hitherto incredible loads over frozen lakes and muskeg. Prospecting ventures that before had demanded years were now cut to months and even days. In conjunction with the continuing policy of the Ontario Government that built the T. & N. O. Railway, the airplane and caterpillar tractor ushered in a period of exploration, discovery and development such as Ontario's northland had never before experienced.

The beginning of Chairman Magrath's tenure of office coincided with an increasing acceleration of industrial production throughout North America and with an awakening awareness of the vastness of Northern Ontario's undeveloped natural resources. As a result, Ontario Hydro's second Chairman faced a highly ironical situation. Less than five years before, his predecessor had been subjected to an almost inquisitorial inquiry to justify the construction, at a seemingly wanton cost to the public treasury, of the Queenston-Chippawa GS and the Cameron Falls development on the Nipigon River, the former to produce an allegedly unneeded and unusable 540,000-hp of electric power and the other called a "white elephant." But now, with the peak load crowding 820,000-hp, Queenston-Chippawa's huge block of power was not only proving insufficient to meet the demand, but so in addition were some 70,000-hp obtained from the numerous lesser generating stations in the Georgian Bay, Trent Valley, and Eastern Ontario Systems. Built or acquired to serve local needs, these minor installations were also subject to lessened stream flow during the autumn and winter months when the power load was at its peak.

More power was urgently needed, but the baffling problem was where to get it. Hydraulic resources regarded as inexhaustible a decade earlier were now found for one reason or another to be strictly limited. Northern Ontario, it was true, held great reserves of power, but these were so far from the populated areas of the province that economic and technical considerations limited their use. In "Old" or Southern Ontario, Hydro, by 1925, was buying power from no fewer than twelve private and public producers, among them the Canadian Niagara Power Company at Niagara Falls, the Hamilton Cataract Power Company at DeCew Falls, the Ottawa and Hull Power and Manufacturing Company on the Ottawa, and the Canadian General Electric Company at Peterborough. Orillia, Bracebridge and Fenelon Falls were the municipal suppliers.

Ontario's greatest power potential, of course, lay in her boundary waters: the Ottawa River, the Niagara River and the international section of the mighty St. Lawrence. But the waters of these rivers were shared; in the case of the Ottawa with the Province of Quebec, and of the others with the United States. Complicating power development

still further was the Canadian canal system on the St. Lawrence operated by the Dominion Department of Railways and Canals; the Georgian Bay Ship Canal Charter prevented provincial action on the
Ottawa River until it was rescinded in 1926.

First projected as early as the eighteenth century, Canada's St. Lawrence canals were inextricably connected with any proposal for power
development of the river. Navigation and electricity were indivisible
and it is impossible to determine which finally had the greater influence
on the eventual development of the St. Lawrence Waterway, although
power requirements certainly provided the initiative.

The joint interest of Canada and the United States in the St. Lawrence dates from the Treaty of Paris in 1783 between Great Britain
and the United States which established the International Boundary
between Ontario and the U.S. along the forty-fifth parallel and
the middle of the River and the Great Lakes. Even before that date
shallow canals of two-feet draft had been built around the Lachine
rapids and between Lake St. Louis and Lake St. Francis above Montreal. In 1821, further steps to improve navigation on the St. Lawrence
to compete with the projected Erie Canal were undertaken by the Provincial Government of Upper Canada, and by 1849 a series of canals
with nine-foot depth had been constructed, with Imperial financial aid,
from tide-water to Lake Erie. Between 1885 and 1902, the system was
again improved to provide locks of 270 feet by 45 feet dimension with
14 feet depth over the sills.

These navigational works were Canadian and wholly unilateral. Following the Treaty of Paris, joint action on the St. Lawrence-Great
Lakes boundary was confined to the Rush-Bagot Convention of 1817,
which limited use of armed vessels on the Great Lakes, and to the
Reciprocity Agreement of 1854, which provided for the reciprocal
use by both countries of all navigational improvements. The next coöperative action was that already mentioned: the informal understanding between Lord Dufferin and Governor Robinson which brought into
being the New York State Park Reservation on the American side of
Niagara Falls and the Queen Victoria Park on the Canadian. A decade
later, in 1895, the creation of an International Waterways Commission
was proposed, to consist of three Canadian and three American members. Finally established in 1905 through the efforts of President
Theodore Roosevelt, the six-man body was charged with the task of
examining and reporting on the condition and the use of boundary
waters common to both countries. Among the subjects discussed were
power rights, irrigation, marine-traffic regulations, stream diversion,
conservation and the maintenance of water levels. But differences in

interpretation and restricted terms of reference, particularly with respect to the diversion of Niagara water for power purposes, soon led the Commission to suggest the establishment of a new and permanent body with judicial powers. After three years of negotiations, conducted for Canada by James (later Lord) Bryce, the British Ambassador to the United States, and for the United States by Elihu Root, the Secretary of State, the Boundaries Waterways Treaty was signed in Washington in January 1909 and proclaimed May 13, 1910. Article XII of the treaty provided for the establishment of a permanent International Joint Commission of six members, three from Canada and three from the United States. The first meeting of the Commission was held in Washington in January 1912 under the joint chairmanship of Th. Chase-Casgrain for Canada and James A. Tawney for the United States.

The Joint Commission is a unique international body in that it was, so far as is known, the first authority anywhere to which two countries had ever delegated a measure of sovereign power. One of its effects was to focus public attention on the magnificent dream of a St. Lawrence waterway by which ocean vessels might some day sail to the interior of the continent. That dream, we know today, was to be delayed for forty years, but so great was the belief in its speedy realization that in 1913 the Dominion Government, acting independently, undertook to provide the first grand link in such a waterway by the construction of a new Welland Canal, between Lakes Ontario and Erie, which rivalled the as-yet-unfinished Panama Canal in many of its facilities and some of its dimensions. At the same time the City of Toronto in coöperation with the Dominion Government embarked on a $30,000,-000 project for the deepening and improvement of the Toronto harbor.

That same year, 1913, Ontario Hydro began an investigation of the great river's hydro-electric potential in connection with its projected Queenston-Chippawa development. The next year, on March 14, the monster radial delegation to Ottawa heard Adam Beck urge that the Federal Government join with the United States in the development of the St. Lawrence for power and navigation purposes as a fitting memorial of a hundred years of peace between the two countries. From that time until his death, Beck remained the foremost advocate of the development.

The gigantic project remained quiescent throughout World War I but, in 1919, Ontario Hydro's first Chairman told the Great Lakes Waterways Conference that the entire costs of a deep waterway from the ocean to the Lakes could be found through the accompanying power development, which, he predicted, would give Ontario the cheapest light, power and transportation in the world. Two years later, Ontario Hydro submitted to the International Joint Commission a proposal based on its separate surveys and research for developing the international power resources and the navigation of the St. Lawrence River. The plan provided for the production of 1,635,000-hp of electricity at a cost of $154,925,415 to be borne jointly by Canada and the United States. The report supported Beck's earlier statements that the entire costs of the undertakings could be paid for by the sale of power in a reasonably short period.

The International Joint Commission next appointed a Joint Board of Engineers headed by two eminent engineers, W. A. Bowden, Chief Engineer, Department of Railways, Canada, and Colonel Wooten, Corps of Engineers, U.S. Army, to study Ontario Hydro's submission. After a year of intensive investigation they reported that it would be entirely feasible to provide a passage for ocean-going ships from Montreal to Lake Ontario, and estimated that 1,500,000-hp of electricity could be generated at a cost of $252,728,000. They were of the opinion also, as were Ontario Hydro's engineers, that the whole cost of a deep waterway and power development could be paid off in a reasonable number of years from the sale of power alone. This report was presented to the International Joint Commission on January 24, 1921. It became the prototype and exemplar for other reports made later by other engineers, and led in 1924 to the creation of a Joint Board of Engineers, to which Canada and the United States each appointed three members, to report on the project more exhaustively. The 1924 letter of transmittal to the Provincial Government accompanying Ontario Hydro's annual report contained the statement: "During the past year, special efforts have again been made by the Provincial Government, by the Municipalities of the Province, and by the Hydro-Electric Power Commission to secure permission to commence construction of the power development works in the international portion of the St. Lawrence River. Delay in securing the permission sought must result in accentuating the power shortage that is rapidly materializing."

Other investigations and reports were made, but formidable opposition to the plan had sprung up from various sources both in Canada and the United States. Premier Taschereau of Quebec was opposed because he believed that it would injure Quebec's maritime interests, despite the assurances of Premier Ferguson of Ontario to the contrary. Prime Minister King in Ottawa watched, and preferred to wait. Private-power interests on both sides of the border were determined that the St. Lawrence should not fall under the control of public ownership, while railway interests, joined by the Port of New York and other cities on the Atlantic seaboard, were opposed to a deep waterway for obvious reasons.

Of course, no one knew better than Charles Magrath, Chairman of the Canadian Section of the International Joint Commission from 1914, the formidable nature of the obstacles that lay in the way of making use of the wasted resources of the St. Lawrence in the forseeable future. As Chairman of Ontario Hydro from 1925 he was also anxiously aware of the pressing need for tapping new supplies of electric energy if Hydro was to continue to feed the ravenous market it had helped to bring into being. Ontario Hydro's engineering staff had also plotted the exigency, and before his death Sir Adam Beck had taken tentative exploratory steps to meet it. Upon assuming office Magrath picked up the same trail, which led to the neighboring province of Quebec.

The hydro-electric development of Quebec's great water powers actually antedated by some years the construction of the first large generating stations on the Canadian side of the Niagara River. As early as 1898 the Royal Electric Company of Montreal had a 20,000-hp plant in operation on the Richelieu River at Chambly, transmitting power sixteen miles to Montreal. The Shawinigan Company installed a plant on the St. Maurice River from which by 1903 electricity was being transmitted more than eighty miles to Montreal. The development of other sites on the St. Lawrence and its tributaries and on the Ottawa River was undertaken before and after World War I to provide power for isolated industries throughout the province, and to serve the immensely profitable metropolitan market of Montreal. Unlike Ontario, however, where Hydro's policy of service-at-cost and the greater dispersion of industry had provided a consistently expanding market, Quebec, where industry was largely concentrated in the Montreal area and where power development had been left to private enterprise, found itself in the mid-twenties with an excess potential supply of power. The existence of such a surplus, coupled with the restrictive charter of the Georgian Bay Canal Company, precluded for the time being any joint development of the boundary waters of the Ottawa River by Ontario Hydro and any coöperating Quebec agency, whether privately or provincially owned.

With a shortage of power in Ontario and surplus of power in Quebec, a mutually beneficial solution of Hydro's problem was at hand, which found expression in an announcement by the Commission stating in part:

> Recognizing the needs for the future the Commission is employing every means within its power to provide for further

development in the most economical manner of the undeveloped provincial water power sites, so that hydro-electric energy may be available at a minimum cost as and when needed to meet the requirements of the public. This policy of public ownership and development does not preclude the Commission from purchasing electrical power when it finds it good business to do so, and such special purchases of power are in no way to be regarded as a departure from general policy of the province with respect to its hydro-electric resources.

The way was now paved for the negotiation of contracts for supplies of energy with large Quebec producers: the Gatineau Power Company, a subsidiary of the International Paper Company, which was building dams and generating facilities on the Gatineau River near the national capital; the Maclaren-Quebec Power Company with plants on the Lièvre River; and the Beauharnois Light, Heat and Power Company, with a station on the St. Lawrence below the International section of the river, built to serve the metropolitan market of Montreal. In addition, Ontario Hydro and the Ottawa Valley Power Company of Quebec were jointly engaged in the Chats Falls development on the Ottawa River. The first and largest contract was signed in 1926 with the Gatineau Company for the delivery of an initial 80,000-hp in 1928; those with Beauharnois and Maclaren-Quebec in 1929 and 1930.

To utilize the power purchased from the Gatineau Power Company it was necessary to construct a 220,000 volt—later 230,000 volt—transmission line from Paugan Falls on the Gatineau River, 230 miles to a great new transformer station to be built at Leaside on the eastern outskirts of Toronto. Both undertakings imposed novel pioneering demands on Ontario Hydro's engineers. The transmission line carried a higher voltage and was longer than any yet projected in Canada, and the Leaside TS required a greater step-down of voltage than was contemplated anywhere.

The Gatineau line was planned to run in a direct diagonal across Ontario from the region of the City of Ottawa to the City of Toronto, traversing in turn the farmlands of the Ottawa Valley, the rugged, lake-spattered terrain of the Shield, and the farming counties running back from the shore of Lake Ontario. In view of the distance and the rugged kind of country involved, conventional ground surveys would have entailed an expenditure of time and energy which at such a critical period was simply not available. In lieu thereof, it was decided to adopt what was then still a virtually new surveying technique by the use of aerial photographs. Both oblique and vertical photographs were made from a height of 10,000 feet and from these the route of the line was determined and tower locations spotted with stereoscopic viewers. Thus prepared, ground survey parties were able to lay out the line and locate the towers in a fraction of the time that otherwise would have been required. Begun in 1927, the line was completed a year later and in 1929 work was started on the longer 330-mile line from Beauharnois to Leaside.

At the time of its construction, the Leaside TS was the largest in the British Commonwealth and one of the largest on the American continent. Among its innovations were the outdoor installation of certain components which hitherto had been always enclosed within roofed walls, and circuit breakers which were physically larger and had a greater rupturing capacity than any manufactured at the time. Interconnected with the Davenport TS at Toronto, the Leaside TS served as the connecting point between the generating plants on the Gatineau and the Niagara Rivers. Construction was begun on January 13, 1928, and on October 31 Premier Ferguson pressed the silver button which inaugurated the exchange of energy generated at points more than three hundred miles apart.

Pending the delivery of additional blocks of Quebec power, efforts were made everywhere throughout Ontario Hydro's systems to improve transmission and increase production. Important among such developments was the consolidation of the Central Ontario, Trent, St. Lawrence and Rideau Systems to provide an interconnected network which extended from Whitby to Cornwall and from Lindsay to Ottawa. This was made possible by Ontario Hydro's purchase from the Province in 1928, for $15,173,235, of the several eastern Ontario properties which were acquired by the Province in 1916 and afterward managed by Ontario Hydro as trustee. When completed, the new Eastern Ontario System equalled that of the Niagara in area.

During the next year, ten generating stations were added by purchase or construction and, as the decade ended, a tenth generating unit was ordered for the Queenston-Chippawa GS, and work begun on the final extension of the powerhouse. The third and fourth banks of transformers and synchronous condensers were ordered for the Leaside TS and the westerly half of the second circuit of the Gatineau-Toronto line was put in operation. Extensive improvements were also completed to improve voltage conditions in the western part of the Niagara System, and additional blocks of power were taken from the Gatineau Power Company in advance of the contractual date. Projected power purchases from Maclaren-Quebec and Beauharnois were also increased, in the first instance to 125,000-hp, and in the second, to 250,000.

The year 1929 also marked the conclusion of negotiations for the purchase of several private power systems; the joint development with the Ottawa Power Company of Chats Falls on the Ottawa River below Arnprior, and the purchase of 100,000-hp from the Abitibi Canyon development of the Ontario Power Service Corporation in Northern Ontario. The most important of the foregoing purchases were the Dominion Power and Transmission Company, with its pioneer generating station at DeCew Falls in the Niagara Peninsula, and transmission lines serving the area from Port Colborne and St. Catharines to Hamilton, Brantford and Oakville; and the generating stations and lines of the M. J. O'Brien interests on the Madawaska River, serving such Ottawa Valley communities as Renfrew and Arnprior. Negotiations between Ontario Hydro and Dominion Power were first initiated in 1917 and finally concluded at a price of $21,000,000, which was later investigated by a Royal Commission of inquiry, as was the O'Brien purchase. The agreement for the development of Chats Falls came at the conclusion of prolonged negotiations between Ontario Hydro, the Province of Ontario, the Province of Quebec, and the Quebec Streams Commission. The Abitibi agreement was made necessary by the demand for additional power, not then locally available, in the Sudbury mining basin between North Bay and Sault Ste. Marie.

New generating capacity added in 1929 from Commission-owned plants totalled 20,509-hp and power purchases over the period 1926 to 1930 for immediate and future delivery amounted to 891,000-hp. These contracts, together with another then under negotiation, were designed to meet requirements to the year 1937. In 1929 peak loads on all Ontario Hydro's systems, excluding exported power, were 8.1 per cent above those of 1928. The curve of Hydro's progress was still rising.

EXPANSION AND CONSOLIDATION

20

In its unremitting search for power, Ontario Hydro had recourse to other negotiations which, like its submission respecting the St. Lawrence to the International Joint Commission and the Federal Government at Ottawa, ended in frustration. In 1927 Ontario Hydro's three generating stations on the Niagara River, together with that of the Canadian Niagara Power Company, were using the 36,000 cubic feet per second provided by the Boundary Waters Treaty between Canada and the United States, signed in 1909. In December 1927 a special board appointed by the Canadian and United States governments made an interim report to Ottawa and Washington with the suggestion that "temporary increase (of allowed water) be made experimentally, under Government supervision, in conjunction with the extension of remedial works (submerged weirs above the Falls) until a satisfactory allowance could be determined." An agreement permitting construction of such works and the temporary diversion of an extra 10,000 cubic feet on either side of the river was signed at Ottawa but was rejected in January 1930 by the Foreign Relations Committee of the United States Senate.

Three months later, in March 1930, the effect of this reverse was mitigated when the New York State Legislature created the St. Lawrence Power Development Commission, later known as the New York State Power Authority. There followed negotiations in which all the interested parties were represented—the Canadian and American Governments, the Province of Ontario and the State of New York, the Hydro-Electric Power Commission of Ontario and the New York State Power Authority. As a result, the St. Lawrence Deep Waterways Treaty was signed by representatives of the two countries in 1932, providing for the coöperative development of the resources of the St. Lawrence. Ratification of the Treaty by the Canadian Parliament quickly followed. Despite the strong representations of President Hoover, and

180

the earlier endorsations of Presidents Harding and Coolidge and of Franklin D. Roosevelt as Governor of New York State, the Treaty failed to win ratification in the United States Senate, to remain dormant for another nine years.

But while the search for power, including the projected development of the St. Lawrence, was a principal concern of Ontario Hydro during the latter part of the 1920's, it was by no means the only, nor even the most pressing concern. Of paramount importance were the responsibilities of the Commission, its staff and organization in the day-by-day operation of a giant utility system whose magnitude and complexity were growing constantly. Among such responsibilities were the daily production and transmission of electrical energy to expanding areas larger already than those comprised in any electrical utility system elsewhere; the constant improvement of equipment and techniques of operation; and provision for future needs through construction and long-term planning. Primarily involved were the operating, engineering and research departments, but ancillary to these were others equally significant—those having to do with maintenance, municipal liaison and supervision, rural electrification, inspection, records, accounting and billing. By the mid-1920's, many of these departmentalized activities required more employees than had sufficed for the entire organization only a few years earlier. Underlying all Hydro activities of course were the Commission's relations with the Provincial Government, from whose legislative enactments Hydro derived both its powers and obligations.

In the course of more than twenty years there had come into being a complicated body of statutory law to which modifications and amendments were made at every session of the Provincial Legislature. Many of these amendments, as Premier Whitney was at pains to point out early in Hydro's career, were enacted to clarify and improve existing statutes, and others simply to give legal status to agreements entered into by Ontario Hydro with municipalities, private power companies and individuals. Although a great number of such amendments were relatively unimportant, being of a routine nature, others had great significance. It follows, therefore, that much of the body of law that governed Hydro's operations was initially tentative and experimental, while some of it was perhaps sheer improvisation.

Reviewing the first two decades of the public-ownership enterprise, it is difficult to see how it could have been otherwise. Unlike the closing years of the period, when Ontario's unprecedented industrial expansion more than justified both the imaginative daring and the exceptionally high costs of the Queenston-Chippawa development, much of the time and energy of those who administered Hydro was of necessity devoted to the political struggle to get the enterprise established and to ensure its growth and survival. To the novelty of the socially inspired enterprise itself was added the novelty of the scientific-technical field in which it was engaged. During at least the first decade of Hydro, the techniques of electrical development were still in such a state of flux that textbooks were of little use: instead they had to be learned at the dam sites, beside the generators, along the transmission lines and in the transformer stations. From Hydro's inception, therefore, two pioneering efforts, the one socio-political and the other technical and economic, had gone forward hand-in-hand, each engaged in its own area to find empirical solutions of problems for which no precedents existed, each requiring legal authorization of decisions made and agreements entered into without benefit of existing statutory enactment.

Under the circumstances it is hardly surprising that the Magrath administration, along with its other multitudinous activities, should have been so largely concerned with legal, financial and fiscal matters. The crusade was over; Hydro had been taken out of politics, and the time had come to manage the municipal coöperative as the gigantic business it had actually become.

Even before 1926 the powers of the Commission were significantly extended. In 1924, for example, all previous legislation governing electrical works, installations and facilities was repealed and Section 17 of the Power Commission Act of that year substituted. The Act empowered the Commission to make rules and regulations with respect to "works and matters used or to be used in the generation, transformation, transmission, distribution, supply or utilization of electrical power or energy in Ontario," thus providing Ontario Hydro with sweeping authority. The same Act gave the Commission authority to unite any two systems in one system and to join to it hitherto independent municipalities. The Act was also amended to increase the maximum payable to the Chairman and other members of the Commission from $15,000 to $45,000. While this sum was not yet commensurate with the salaries paid in private industry for posts of equivalent responsibility, the 200 per cent increase reflected a changing attitude toward the Commission both in the minds of the Government and the public whom Hydro had served, sometimes at personal financial sacrifice. Legislation passed in 1919 and amended in 1925, pertaining to Ontario Hydro's pension fund established in 1923, was a further indication of Hydro's institutional maturity.

In 1926 the Power Commission Act made provision for the payment to the Province of all funds secured by the Commission on sinking fund account. The schedule called for sums ranging from $1,338,-567 in 1927 to $6,179,317 in 1966. The Act also authorized the Provincial Treasurer to sell securities of a par value of $4,812,000 which had been deposited by the Commission toward the repayment of advances made by the Province. Such advances now totalled $135,-049,183 or, with the reduction above noted, $130,237,183. Investment by Ontario Hydro in power undertakings and electric railways amounted to $203,442,757 and that of the municipalities $74,692,540, totalling $278,135,297. Its revenues for 1926 totalled $20,555,179, of which $565,413 was returned to the municipalities in the 13th billing. For the year, the rate to domestic consumers averaged less than 2 cents per kwh, while rates in the United States, as reported by *The Electrical World* of January 1, 1927, stood at 7.4 cents per kwh for comparable service. In the course of twenty years, Ontario Hydro had already become one of Canada's larger corporations, although its growth was hardly yet begun.

Continuing the legal pattern of this period, the Ontario Legislature in 1927 passed a complete revision and consolidation of the Power Commission Act, which had stood on the statute books since 1906. Prepared with the coöperation of Hydro's legal staff, and contained in Chapter 57 of the Revised Statutes of Ontario for 1927, the Act did nothing to alter the Commission's powers or obligations; it did, however, bring simplification and order to a body of law which in the course of twenty years was tending to become chaotic.

But while the Revised Statutes of 1927 codified the Law affecting Hydro, they fail to reveal other interesting relations between the Commission and the Province. Included in this category are such matters as the financial aid granted for rural electrification and the development of Northern Ontario, and the complex arrangements between the two bodies in connection with the various properties whose ownership, for fiscal purposes, was vested in the Province but which were managed and operated by Ontario Hydro and were to all practical intents and purposes under its sole control.

Of special interest from a fiscal standpoint was the payment by Ontario Hydro to the Province of more than $2,000,000, $1,455,122 of which represented the full settlement of water rentals charged to the Queenston-Chippawa GS to October 1938, and $545,997 the amount owing by the Ontario Power Company and the Electrical Development Company at the time of their transfer to Ontario Hydro ownership in 1924. The arrears, in the second instance, had arisen from the difference in charges based on the "average load" and "minute peak" standard adopted without formal agreement a short time before the purchase. Under the agreement the charge for water rentals was made on the minute-peak maximum reached and held for one minute which became the rate for electrical horsepower for a six-months' period or until exceeded by another minute peak, whereupon the new maximum replaced the first for the remainder of the six-month period. The effect of the payment of such arrears by Ontario Hydro, and hence by the municipalities, was to furnish a better basis for comparison between its operating costs and those of privately owned utility systems in Canada and the United States.

Other developments during 1925-1930 involved changes in the regulations previously embodied in legislative enactments concerning the formulation of advanced standards for electrical wiring, equipment and apparatus, and the adoption of rigid specifications to make these effective. These activities were the culmination of a process of organic growth which had its humble beginnings in the Research Division, set up under H. D. G. Crerar in 1911, and in the establishment of the first laboratory in Toronto in 1912. Out of this had grown testing and research laboratories whose duties embraced: (1) tests and investigations on materials and apparatus received in the storeroom on stock orders or submitted to the laboratories by the various departments of Ontario Hydro or by outside parties; and (2) tests and investigations relating to problems arising from the operation of the system. This in turn had led to the formation of an approvals staff of four men whose numbers were rapidly augmented as the demands for testing were increased. By 1914 laboratory activity had expanded to include a High Tension and General Testing Laboratory, Lamp Testing Laboratory, Meter and Standards Laboratory, and an Illuminating Engineering Laboratory, all of which were financially self-supporting.

Hand-in-hand with this expansion occurred the evolution of the inspection service. Inaugurated in 1913 with the publication of Ontario Hydro's first book of Rules and Regulations, the service had been given mandatory powers, making Hydro the first organization of its kind in the world to introduce electrical safety regulations which were enforceable by law. In 1915 a separate Electrical Inspection Department was established and by 1917 the inspection service had been extended to cover 501 municipalities.

As the new decade opened, the Department was kept increasingly busy because of extensive new building construction everywhere, with a consequent increase in the use of domestic electric appliances, particularly kitchen ranges. Despite the temporary post-war depression that followed in the early 1920's, the number of domestic users of electricity continued to grow. Canadian electrical manufacturers and suppliers were producing new lines of goods or adding to old ones and in the year 1922 alone, submitted nearly two hundred new or improved devices for approval. Local Commissions were also establishing retail merchandise outlets where appliances could be bought on easy terms, and electrically equipped homes were on exhibition in many cities of the province. All this increased the demand for supervision, which was also extended to cover advertisements in the daily and technical press, and electrical installations at fairs and exhibitions.

The work load was increased further by the rapid extension of rural lines, stimulated by grants-in-aid; but by 1922, after ten years of operation, inspection procedures, now divided into thirty-two districts, had become standardized and could be carried out expeditiously.

Meanwhile, the Hydro Approvals Laboratory, working in close co-öperation with the Underwriters' Laboratories of New York and Chicago, and with the Canadian Engineering Standards Association, had continued its work of establishing new standards for all kinds of electrical installations and appliances.

The Inspection Department was given authority to enforce Ontario Hydro's Rules and Regulations throughout the province, including municipalities not in the Hydro partnership. At the time it was instituted, this was probably the most comprehensive and effective system existing anywhere, and undoubtedly the simplest, since its effectiveness lay in the ability of the individual inspector to refuse to furnish a permit for the turning on of energy for any installation that failed to obtain his approval. As a corollary to Ontario Hydro's pioneer work in the fields of standardization and inspection, the Canadian Electrical Code was completed in June 1927 by the Canadian Standards Association, and was soon adopted not only for Ontario but for the provinces of Quebec, British Columbia, Nova Scotia and Saskatchewan.

Along with the crystallization of administration practices, this period was marked by a steady growth in rural electrification as well as important changes in its pattern of distribution. In 1924, for example, when provincial grants covering 50 per cent of the construction costs of rural lines, authorized by the Rural Hydro-Electric Distribution Act of 1921, were extended to include transformers and secondary equipment, 1,205 miles of primary lines had been built, and these served 20,605 consumers, or approximately 17 to the mile. Four years later the comparative figures stood at 3,790 miles of line and 31,000 consumers, approximately eight to the mile, representing the extension of this type of service, possibly the most important from a social and cultural standpoint, to areas of less concentrated population. That same year, 1928, 929 miles of new construction and 6,000 new customers were added. Ontario Hydro was now operating in 122 rural districts comprising 211 townships, and the aggregate peak load had reached 16,890-hp, as against 3,514-hp in 1923. The total investment in rural electrification including government subsidies was now approximately $7,300,000.

Referring to the increased tempo of rural light and power expansion, Chairman Magrath stated in his 1928 report: "The endeavor of the Commission is not the concentration of industry at a few large power sites, but rather the broader policy of making as widespread distribution of electrical energy as is economically possible."

The words "economically possible" take on added significance in their historical context. They were not merely a reference to the financial ability to provide rural electrification but to the demands being made upon Ontario's capital resources by the industrial growth then taking place. Not only was the expansion of Hydro itself proceeding at an ever-accelerating pace, but the Provincial government was also committed to the investment of vast capital sums for the development of Northern Ontario and for the modernization of Ontario's highway system. In such a context the humanitarian goals of rural electrification, however desirable, could be approached only as fast as money, labor and materials were available.

There was, however, another inhibiting factor: the attitude of many potential rural customers themselves. For a few years after World War I, introducing electricity to the farmer was largely a problem of salesmanship. At the time manpower was plentiful, and the wages paid to hired help relatively low, so that the farmer regarded a long-term contract with Ontario Hydro as somewhat in the nature of a mortgage on his farm. He was unable to see how, apart from lighting, he could make any economical use of electricity. Soon, however, as the result of agricultural mechanization and industrial development, it became increasingly difficult for the farmer to keep men on the farm, while to those who were willing to remain he was forced to pay higher wages.

Before 1939 practically every Ontario Hydro engineer working in the rural field had to assume the role of salesman. Meetings were held in all the rural districts to stimulate interest in the use of electricity on the farm and increase the number of customers. It had to be demonstrated that if the ratio of customers per mile of line could be substantially increased the rates to farmers would be greatly reduced. It is on record that an Ontario Hydro engineer on one occasion determined that the only way he could get his third customer—a woman poultry farmer—to sign up was by agreeing to buy from her thirty dozen eggs every month so that she would be assured of enough money to pay for the costs to be incurred. He bought the eggs and sold them to his friends until such time as Ontario Hydro's new customer found electricity such a boon that the engineer was able finally to cease his entrepreneurial endeavor and return to a more balanced diet.

It will be recalled that, from the inception of Ontario Hydro, rural electrification and all that it signified in terms of social amenities and improved standards of living had been one of the cherished goals of public-ownership supporters. It was not until 1912, however, that effective steps were taken to attain that objective, and not until 1921, when the decision was made to provide government subsidies, that substantial progress was realized. During the ensuing seven years, the rate of growth was so accelerated as to draw world-wide attention to Hydro's achievement. It is interesting to note that the United States Rural Electrical Administration, inspired largely by the success of the Ontario experiment, was not inaugurated until 1935, and that the U.S. Rural Electrification Act, which followed closely the Ontario legislation, was not passed by Congress until 1936. The principal difference between the American and Canadian statutes lay in the absence in the former of any provision for direct grants for construction, although long-term loans were made available for that purpose and for the installation of wiring and electrical apparatus. While it is true that in some countries which had a heavy concentration of farming population, such as France, Germany, Holland, Sweden and New Zealand, rural electrification was in advance of Ontario; nevertheless, in the light of the greater distances and sparser population involved, its achievement should have given great satisfaction to those responsible. Yet the rate of progress was deemed unsatisfactory.

The stumbling block lay in the meagre cash resources of so many farmers. Even when electricity was brought to their doorsteps, they were unable to finance wiring or the purchase of electrical appliances without which the boon of low-cost energy was meaningless. To remedy this situation, the Legislation in 1930 passed the Rural Power District Loans Act, which provided loans up to $1,000 repayable with interest in twenty years, for such purposes as wiring in farmhouses and barns and the installation of motors and other labor-saving devices.

The Act was put into operation early the following year, when it was turned over to the Commission for administration. Interest was fixed at 6 per cent payable quarterly along with payments of the principal. To finance these loans a sum of $2,000,000 was set aside out of the Consolidated Revenue Fund. Out of this fund also were paid any deficits in new rural districts where the maximum service charge was insufficient to meet costs, but such deficit payments were to be repaid out of any surplus arising later from the maximum service charge in that rural district. Another enactment of 1930 was the Rural Power District Service Charges Act, which gave the Commission power to regulate rural service charges so that relatively high charges would not be set for the first customers in a rural power district to the benefit of customers taking contracts later.

Hydro's tremendous expansion of the 1920's, beginning with the first surge of energy from "the Largest Hydro-Electric Plant in the World" at Queenston in December 1921 and culminating with the obligatory purchase of even larger blocks of power from private sources outside the Province was, of course, one of the many revolutionary concomitants of the industrial boom of the period. When the death knell of that boom was sounded in Wall Street on October 24, 1929, Ontario Hydro was in a stronger position than it had ever been in its short but tumultuous history. In the relative political tranquillity of the few preceding years, phenomenal progress had been made in every area of the Commission's activity and for the first time it was able to look forward to sufficient future reserves of power to meet the annual increment in demand. Both management and organization had been strengthened and made more efficient. The financial position *vis-a-vis* the Provincial Government had been greatly improved by the progressive amortization of its capital advances and by the setting up of stronger reserves by both Ontario Hydro and the municipal commissions. Entering a depression which was to prove as unprecedented as the boom that preceded it, Hydro's position seemed as impregnable as human ability and integrity could make it.

21

In December 1930 G. Howard Ferguson resigned as Premier of Ontario to become Canadian High Commissioner to London, and was succeeded by the Hon. George S. Henry, a member of the Legislative Assembly since 1913 and a Cabinet minister in both the Hearst and Ferguson Governments. Two months later, in February 1931, the resignation of Charles A. Magrath as Chairman of Ontario Hydro became effective and the Hon. J. R. Cooke, a member of the Commission since 1923, was appointed in his stead. C. Alfred Maguire continued as the nominee of the OMEA and in June 1931 the Rt. Hon. Arthur Meighen was named to fill the vacant post on the three-man board. W. W. Pope and Fred Gaby, both of whom had been with Ontario Hydro since its early days, continued as Secretary and Chief Engineer respectively.

Signing Ontario Hydro's annual report for 1930 J. R. Cooke, as Acting Chairman, paid this tribute to his predecessor:

> Perhaps the outstanding effort which he [Magrath] put forth was in connection with the provision for future supplies of electrical energy to meet the needs of Ontario municipalities. . . . The production of additional hydro-electric power from waters owned by the Province was most carefully considered and furthered wherever economically possible. Definite attention was given to the desirability or otherwise of developing steam-electric power, and the possibilities of purchasing power in large blocks to be supplied from sites owned or controlled by private organizations were examined.

It was Magrath's acceptance of the last solution, specifically the purchase of 791,000-hp from Gatineau, Maclaren-Quebec and Beauharnois at $15 per horsepower, that provoked the most heated criticism during the Cooke administration.

189

As was to be decisively shown in a few years' time, the Quebec power contracts made between 1926 and 1934 represented industrial statesmanship of the highest order. Without them, even by December 1934 when the peak demand for primary power exceeded the dependable capacity of the Commission's generating stations in the Niagara System, there would have been a shortage of power. Notwithstanding the great significance of this achievement, Magrath's contributions in other ways were equally noteworthy. Under his chairmanship there was a strengthening in the financial position of the Hydro enterprise, evidenced by the fact that the aggregate reserves of the Commission and the municipalities were increased by more than $57,000,000 to nearly $104,000,000, while at the same time the capital invested in the various systems and municipal undertakings increased by almost $73,000,-000 to a total of over $359,600,000. The retiring Chairman was able to turn over to his successor a greatly improved and expanded physical plant. In 1930 this embraced the following five systems:

1: *The Niagara System,* in which the Niagara generating plants supplied the area between Niagara Falls, Hamilton and Toronto on the east, and Windsor, Sarnia and Goderich on the west—approximately the same section of the province as that originally served by Hydro from Niagara. The Niagara plants included those formerly owned by the Ontario Power Company and the Toronto Power Company, as well as the great Queenston-Chippawa development. To them in 1931 were added the plant and transmission lines of the Dominion Power and Transmission Company, serving the area embracing Port Colborne, Hamilton, and Brantford. Energy from these sources was supplemented by power transmitted from the Gatineau River in Quebec to the Leaside TS. Into the Niagara System had also been incorporated the Essex County and Thorold Systems, the Ontario Transmission Company, and the Toronto and Niagara Power Company.

2: *The Georgian Bay System,* a consolidation of four systems—the Severn, Wasdell, Eugenia, and Muskoka—to which had been added in 1930 the Foshay interests in Bruce County by purchase from the Public Utilities Consolidated Corporation of Minneapolis. The System supplied that part of Ontario surrounding the southern end of Georgian Bay and Lake Simcoe as far north as Huntsville in the Lake of Bays district. It was served by eleven generating stations, including Wasdell Falls and Big Chute on the Severn River; the Eugenia on the Beaver River; and the South

Falls and Hanna Chute developments on the Muskoka River. Additional power was obtained by interconnection with the Niagara System and by purchase from plants at Orillia, Port McNicoll and Owen Sound.

3: *The Eastern Ontario System*, which served the entire eastern part of the province and was, again, a consolidation of a number of smaller systems: the Central Ontario and Trent, the Rideau, and the Ottawa and St. Lawrence Systems, to which was later added the Madawaska System. Of these, the Ottawa System served the City of Ottawa and an extensive rural power district in the vicinity, purchasing its power from the Ottawa and Hull Power Manufacturing Company whose plants were at the Chaudière Falls on the Ottawa River adjacent to the city. The Rideau System served the area in the neighborhood of Carleton Place, Perth and Smith's Falls from plants located on the Mississippi River at Carleton Place and High Falls. Additional power was bought from the Rideau Power Company at Merrickville on the Rideau River. The St. Lawrence System served the district immediately north of the St. Lawrence River, with power generated at Cedar Rapids on the St. Lawrence and bought from the Cedar Rapids Power Company. The remaining unit of the Eastern Ontario System was the Central Ontario and Trent, serving the district bordering on the north shore of Lake Ontario between the Niagara and Georgian Bay Systems to the west and the Rideau and St. Lawrence Systems to the east. The nucleus of the Central Ontario and Trent System had been the group of properties formerly controlled by the Electric Power Company and operated by it through the agency of twenty-two subsidiary companies until 1916, when they were purchased by the Province and turned over to Ontario Hydro for operation as trustee until 1928, when the Commission bought them. By 1930 a substantial amount of power was being purchased from the Gatineau Power Company in Quebec for all these customers. Increased consumption led to an enlargement of the system, for which local power was obtained from nine generating stations on the Trent and Otonabee Rivers.

4: *The Thunder Bay System*, which served the district at the head of Lake Superior and included the twin cities of Port Arthur and Fort William. Power was also supplied to the village of Nipigon and to pulp-and-paper companies in the area from developments at Cameron Falls and Alexander Landing on the Nipigon River. A dam had been built at Virgin Falls below the outlet of Lake Nipigon to provide storage and regulate the river's flow.

5: *The Northern Ontario System*, which included the Nipissing and Sudbury districts lying northeast of the French River and Lake Huron, and the Patricia district in the far northwest section of the province. Nipissing was served by three generating stations at Nipissing, Bingham Chute and Elliot Chute, all on the South River, with Elliot Chute under remote control from Bingham Chute. The Sudbury district was also served by three generating stations which were located on the Wanapitei River. When these installations were no longer adequate to supply the burgeoning mining industry of the Sudbury Basin, Ontario Hydro in 1930 undertook construction of a 110,000-volt transmission line to connect the International Nickel Company's plant at Copper Cliff with Hunta, 189 miles to the north, where 100,000-hp would be delivered by the Ontario Power Service Corporation from a new generating station being built at the Abitibi Canyon.

The Patricia district occupied the remote area of the province lying beyond the Height of Land and adjacent to the Manitoba border. There, on December 25, 1929, a generating station at Ear Falls on the English River began operation and two months later was furnishing power to a single gold mine in the Red Lake district still farther to the north. In many respects this was one of Ontario Hydro's most remarkable engineering achievements, since the area was so remote that it presented almost insurmountable difficulties for transporting men, supplies and materials. During the fall and spring, when northern travel, even by airplane, becomes virtually impossible, the only communication with the outside world was by short-wave radio, then still in an experimental stage. Nevertheless, a dam had been completed early in 1929 to create a vast new interior storage basin in Lac Seul and to control the flow of the English River upon which various power sites were located between Lac Seul and Lake Winnipeg; a powerhouse was later constructed at the dam, massive generating machinery was installed, and a transmission line was erected through a wilderness that hitherto had known only the trapper and prospector.

In addition to the widely ramified and highly interconnected systems existing in February 1931, when J. R. Cooke and his colleague C. A. Maguire took over, various projects were in course of construction or nearing completion which had been undertaken before the depression, anticipating a demand for power. These included the joint development of Chats Falls on the Ottawa River in association with the Ottawa Valley Power Company, a Quebec corporation; the construction of a third 220,000-volt transmission line from that development to Leaside; the integration of the Dominion Power and Transmission Company's facilities with those of the Niagara System; and the installation of a second and third generating unit at the Alexander development on the Thunder Bay System.

This gigantic complex of generating plants, transmission lines, transformer stations, and distributing circuits, which now served 668 coöperating municipalities embracing 26 cities, 94 towns, 251 villages and 297 townships, had been built and was constantly being improved and expanded, to meet anticipated future demands for power based on the Commission's past experience. Without exception, the annual demand for power had risen constantly from 1910, when the first surge of energy reached Berlin (Kitchener), until 1930, the first full year of the Great Depression. In 1931, however, as the wheels of industry slowed down so did the demand for electricity, and for the first time in its 25-year history Ontario Hydro found itself with surplus power on its hands and under contract to pay its Quebec suppliers for large additional blocks of power for which there was no market.

Viewed in retrospect, Ontario Hydro's record for the critical years 1931-1934 is seen to have been much better than it was made to appear at the time. In fact Ontario Hydro, in common with other Canadian producers of hydro-electric energy, generally suffered less than most industries in the country, as the accompanying table shows.

ONTARIO HYDRO'S *Gross Revenues*
from Municipal Utilities and other Power Customers

1930	$32,180,919
1931	31,526,568
1932	28,055,895
1933	27,520,853
1934	29,451,564

As can be seen, the actual decline in revenue from 1930, at the beginning of the depression, to 1934 was approximately 18.6 per cent. Thereafter the familiar pattern of annual growth was resumed, to continue, with a single exception in 1938, until the present day. During the same 1931-1934 period, the total load of power generated and purchased by Ontario Hydro declined from a peak of approximately 5 billion kwh in 1930 to a low of 4.2 billion kwh in 1932, when it, too, resumed the upward climb to reach a new high of 6.4 billion kwh in 1934.

During the same 1932-1935 period, however, Ontario Hydro suffered a series of deficits which totalled approximately $12.5 million, caused in part by the purchase of unsalable power from Quebec and by a decline in exports of power to the United States. Instead of charging this operating loss to the municipalities and hence to domestic consumers, it was met by appropriations from the obsolescence and contingencies reserve. Notwithstanding substantial withdrawals, the combined reserves of both Ontario Hydro and the municipalities were increased from approximately $122 million in 1932 to almost $148.5 million in 1935, from which nearly seven million dollars was repaid the Province on Ontario Hydro's capital account.

The apparent paradox of building reserves and retiring debt while operating at a loss is readily explained by the fact that amounts collected for depreciation and sinking fund exceeded the withdrawals from contingencies reserves. Because of Ontario Hydro's decision not to raise charges in proportion as unit costs increased, there was no falling of domestic consumption, the drop in load being wholly attributable to the decline in industrial use and in export demand. Throughout the depression, the very cheapness of Hydro's product made it one of the few commodities which almost everyone could afford in spite of reduced wages or unemployment. As a result, domestic consumption climbed slightly. To stimulate this trend, further rate reductions were made, and in 1933 a vigorous campaign was launched by Ontario Hydro and the municipal sales organizations for the greater use of water-heaters and other heavy-duty electrical household appliances. So successful was the campaign that it paid for itself from the beginning, through sales of equipment and expansion of domestic consumption.

The same general patterns of sales promotion were followed in industrial and rural operations. Large blocks of reserve power were offered at low rates for export, and to pulp-and-paper companies for generating steam, while rural users received further rate reductions and, in addition, the benefits of the Rural District Loans Act passed in 1930. Although the rate of rural expansion was less than in the closing years of the preceding decade, 2,020 miles of rural lines were built and 15,130 new customers added in the years 1931 to 1933 inclusive. By 1932 major construction on Ontario Hydro's five systems having virtually ceased, and future long-range planning having been brought to a standstill by the refusal of the United States Senate that year to ratify the St. Lawrence Deep Waterways Treaty, it became necessary to make drastic reductions in the construction, designing and drafting personnel.

Paralleling Hydro's remarkable record in maintaining and even adding to its domestic and rural loads were the developments which took place in Northern Ontario during the depression. While the pulp-and-paper industry in general experienced a sharp decline in sales, the opposite was the case with mining, particularly after 1933, when the price of gold was raised from $20.67 to $35 an ounce. A spectacular expansion in precious-metal production followed and again Hydro found itself in the traditional position of having to provide for expanding demand.

Except for its installations on the Nipigon River and the newly constructed Copper Cliff-Hunta transmission line, Hydro's Northern Ontario activities up to this time had been on a relatively small scale, and operated on a different basis from those in the more settled parts of the province where the coöperating municipalities had borne their share of the cost of installations from which they benefited. This had never been possible in the remote and sparsely settled areas of the north, where few organized municipalities existed. In the interests of northern colonization and development, the Provincial Government provided the capital and Ontario Hydro acted as agent for the Government in the engineering and construction, as well as in the administration and technical operation, of the properties.

In 1933, to assure a more rapid development of Ontario's vast northern domain, an area more than seven times larger than Old or Southern Ontario, the Legislature in 1933 amended the Power Commission Act to permit Ontario Hydro to build and operate works in any of the territories of the province: Kenora, Rainy River, Thunder Bay, Cochrane, Algoma, Temiskaming, Sudbury, Nipissing, and Manitoulin Island, all lying roughly north or west of the Mattawa-Nipissing-French River line. The Act provided that Ontario Hydro should finance the undertaking but that deficits would be paid out of the Province's Consolidated Revenue Fund, to be repaid with interest from future surpluses: and that properties built or acquired be held and administered by Ontario Hydro in trust for the Province. To aid further in the development of the northern gold fields, which could not be economically exploited without low-cost electric power, Hydro's rates to all territories other than the first three listed above were set at $32.50 to $35 per horsepower, regardless of the actual costs of production and transmission to the mines.

The first property acquired under the Northern Ontario Properties Agreement was the uncompleted Abitibi power development and transmission system of the Ontario Power Service Corporation. With its inability in 1932 to meet the interest on a twenty million dollar bond issue, the Montreal Trust Company was appointed receiver and manager of the assets of the Corporation on request of the Commission as major bondholder, and Ontario Hydro, acting as agent for the receiver, took over the transmission line system and resumed construction of the critically needed far-northern development. Negotiations between the Province and the bondholders followed, resulting in the purchase and operation of the assets by Ontario Hydro. The purchase received legislative approval at the 1933 session of the Assembly and the first power was delivered from Abitibi Canyon GS to Copper Cliff on May 24 of that year. Despite the speed and efficiency with which work on the Abitibi development was rushed to completion to supply sorely needed power for the expanding mines of the Sudbury Basin, the Abitibi purchase, as will be seen, was to furnish critics of both the Henry Government and Ontario Hydro with some of their most effective ammunition.

The year 1932 was marked by one of the most romantic events in the long story of Hydro's rural electrification. This was the extension of rural service to one of the most picturesque sections of the province, the Manitoulin district, comprising a number of islands, one of them said to be the largest fresh-water island in the world, lying along the north shore of Lake Huron. Known to the Indians as the dwelling place of the good spirit *gitchi-manitou* and the evil spirit *matchi-manitou*, to the early French explorers and fur traders for the sheltered route the islands provided to the *pays d'en haut*, and to generations of hunters and fishermen for the fine sport to be found there, Manitoulin Island itself more recently has been the scene of archaeological discoveries dating human occupancy back 12,000 years.

22

Seen in its historical context, Hydro's depression record reflects unswerving devotion to the principles of public ownership during a period of grave economic demoralization. At no time was service impaired or the efficiency of the system allowed to deteriorate, nor were any expenses incurred that could have been prudently avoided. Notwithstanding such a record of achievement, at no time in Hydro's history was the Commission exposed to such relentless and damaging criticism, most of it politically inspired. In the beginning, charges and criticism were levelled less against the acts of the Cooke-Maguire-Meighen triumvirate than against those of the preceding administration: however, as the political storm rose to hurricane force, any distinction between the two commissions was lost sight of. The most persistent and, as it turned out, the most effective critic was Mitchell F. Hepburn, in 1926, at thirty years of age, the youngest member ever elected to the House of Commons in Ottawa, and from 1930 the leader of the Liberal Party in Ontario.

Hepburn's attacks on Ontario Hydro's administration began in 1931, with charges of incompetence and worse in connection with the 1926-1929 contracts for the purchase of large blocks of power from Quebec producers. Three years later, by mid-1934, when the Provincial election campaign of that year had reached a climax of charges and counter-charges, few aspects of Hydro's policies and operational system had escaped the attentions of politicians in the Legislative Assembly and outside.

That the Legislative Assembly should have evinced a deep interest in Hydro's affairs is not in itself surprising: as the representative of Hydro's ultimate owners, the people of Ontario, the Assembly had a manifest duty to be concerned with the workings of the Commission. What does require explanation, however, is the manner in which Mitchell Hepburn and his followers were able, despite the record and

197

without ever actually proving anything, to discredit Ontario Hydro with the general public by imputing inefficiency, and even corruption, in its management.

In a confused political and social climate, where people everywhere were seeking a scapegoat for a demoralized economy, Ontario Hydro's inability to protect itself adequately against damaging outside criticism appears to lie in a complete failure of communications. It had, in fact, lost touch with the people. This failure, which originated more or less unobtrusively under the chairmanship of C. A. Magrath, was the result of an aristocratic approach to the question of providing information to the public. Magrath, a first-rate civil servant, widely experienced in the administration of government agencies and commissions, but used to working and negotiating in privacy, behind the scenes as it were, regarded all publicity media with disapproval, and treated the press with a reserve bordering on hostility. As Chairman of Ontario Hydro, he submitted a memorandum to the Provincial Government in 1929, containing the following revealing passage in defence of his reluctance to publicize the Commission's activities:

> It does appear to me that the Commission, through its annual report, gives details of its work, even to a greater extent than I think is desirable. We must realize that the Commission is a *commercial organization*, and while there is nothing to hide, so far as we are concerned, it is not fair to those engaged in hydro-electric efforts, some of whom seem to be willing to discredit us with the public. There is not a *private corporation* in this country that could afford to operate on such a basis. (*Italics added.*)

It is true that a body such as the International Joint Commission, over whose Canadian Section Magrath presided as Chairman from 1914 to 1935, might operate most effectively with a minimum of publicity, detached from press conferences, informational and explanatory releases, and other educational methods for fostering an informed public opinion on its affairs; the Joint Commission's activities were too remote from direct public concern to warrant a comprehensive grassroots communications policy. Ontario Hydro, on the other hand, enjoyed a diametrically opposite relationship with the public. It was not and had never been a *commercial* organization, no matter how great the similarities between its management techniques and those of a private corporation.

The Magrath Commission did continue the practice of publishing an Annual Report in book form, but while this document, replete with graphs and statistical tables, serves as a very useful guide for engineers and accountants, its treatment of Hydro's complex of operations and projects is too technical to engage the interest of the general reader.

The danger in neglecting to ensure a well-informed public opinion of Hydro was that it gave its opponents the opportunity to fill the void with unfavorable half-truths and unproven aspersions. That this did not happen during Magrath's tenure of office may be attributed to the flourishing, if unstable, industrial productivity of the twenties, which inhibited even constructive criticism in almost all branches of industrial and commercial enterprise. When the bubble burst in 1929 the general attitude of carefree optimism changed equally swiftly to a mood of angry censure, rampant with an iconoclastic determination to shake to its foundations every theory and institution. Hydro, engrossed in problems of maintaining a balance between supply and demand, was caught off its guard, and Mitchell Frederick (Mitch) Hepburn was both ready and able to make the most of his opportunity.

Hepburn, though Leader of the Liberal Party for the Province of Ontario, retained his seat in the Federal Parliament until a short time before the Provincial elections of 1934. In the meantime, the stage was set for him in the Ontario Legislative Assembly in 1931, by repeated Opposition attempts to discredit the Commission over the signing of the Quebec contracts. In March, a Liberal member, G. A. McQuibban, described as "iniquitous" the fact that 122,000-hp of energy should be exported to the United States while simultaneously a like amount was imported at great cost from Quebec. Premier George S. Henry intervened to remind him that when Ontario Hydro took over certain private companies—specifically, the Ontario Power Company and the Electrical Development Company—it assumed at the same time the contractual obligations of those companies to export a certain amount of power. This reply clearly absolved Ontario Hydro from any blame regarding the unavoidable export of power at a time when the demand in Ontario exceeded the available supply. McQuibban's accusation did, however, introduce into the Legislature the subject of the Quebec contracts. Very shortly, these contracts would require the importation of power in considerably larger quantities, while unforeseeable decreases in industrial demand would leave Hydro faced with a surplus of locally generated power. The public, unaware of the circumstances producing such a seeming paradox, were apt to arrive at wrong conclusions, and were unfortunately encouraged to do so by the many interests opposed to either the contemporary Hydro administration or the entire principle of public ownership.

To refute adverse criticism based on the Quebec power companies' bond prospectuses, the Commission issued a formal statement on May

6, pointing out the irrelevancy of the criticism and emphasizing how unsatisfactory a source of information the prospectuses were in view of the difference between Hydro's purposes and those of profit-seeking corporations. But formal statements inevitably lack the impact of a skilled presentation of the facts to the general public, planned to induce the layman to read it and absorb their import. What Hydro needed was something which Sir Adam Beck had provided apparently by instinct —a close contact with the press, as the best medium for communicating the Ontario Hydro story to its millions of customer-owners.

Hepburn entered the lists at this point, charging in a speech at Milton in May 1931 that the Commission had been corrupted by politics, and claiming that power costs were too high. He followed, while campaigning for the Liberal candidate in the South Wellington by-election, with a demand for an investigation into the Hydro-Beauharnois Power Corporation contract of November 29, 1929. In July 1931 the Provincial Parliament appointed a Special Committee of Inquiry into the Beauharnois contract, and certain evidence provided before this Committee prompted Hepburn, some months later, to demand a Royal Commission to investigate the Beauharnois and other contracts made by Ontario Hydro.

The offensive was renewed at the Annual Meeting of the Ontario Liberal Association in London, October 20-21, when Hepburn asserted that the public had now reached a point where they were questioning the price of their power. Characteristically, he demanded further investigations, this time into the Chats Falls, Abitibi, and Maclaren developments, and decried as exorbitant the purchase price paid by Ontario Hydro for the Dominion Power and Transmission Company. The pressure of adverse criticism continued to harass the Commission, so that even when Chief Engineer Gaby made certain reasoned statements in rebuttal of the charges laid against Ontario Hydro, he was accused of engaging in political controversy.

In its Annual Report for 1931, the Commission detailed the circumstances that had brought about the signing of the contentious Quebec contracts:

> The Commission, at the commencement of the period of depression, had practically no margin between its available supplies and the actual demands. For a year or more following, the Commission, with its provision for power supplies, was merely building up a reasonable reserve in relation to the power actually in use.

How prudent was this policy one example of its results will show. The serious water shortages of that year, 1931, in the Georgian Bay and Eastern Ontario Systems were only prevented from causing heavy power deficiencies by the purchased supply under the Gatineau contract, and by other far-sighted arrangements, including the building of a 15,000-kva transformer station at Kingston, which enabled the Eastern Ontario System to take additional 60-cycle power at a date earlier than that provided in the contract.

The next broadside fired at the Commission occurred at Ottawa in February 1932, when Hepburn demanded to know why the Commission had contracted to pay the Gatineau Power Company in U.S. funds, at a cost to the Province of $600,000 at the current rate of exchange. He declared that the books of this company, as well as those of the Dominion Power and Transmission Company, should be produced for public examination. He named the Abitibi Canyon Power Company, the Chats Falls development and Beauharnois as three companies to which Ontario Hydro, though receiving only part of the power produced, was paying enough annually to take care of more than their various carrying charges.

The effect of Hepburn's accusations appeared to be greatly weakened when Premier Henry presented certain facts reflecting on the sincerity of the Opposition Leader's motives. Addressing the Provincial Parliament of February 18, Henry read excerpts from an anti-public-ownership pamphlet published in Chicago, and matched them word for word with excerpts from Hepburn's speeches, quoting the pamphlet itself to this effect: "In case of a thorough-going anti-municipal-ownership campaign you will find this information indispensable." "The Liberal Leader," Henry continued, "stands up and expresses faith in Hydro and public ownership, and then he seizes a copy of an organ directly built for the purpose of destroying public ownership anywhere."

To put an end, if possible, to this persistent harassment, the Henry Government agreed to the appointment of a Royal Commission to inquire into the more serious allegations of mismanagement besetting the Ontario Hydro Commissioners. Originally it was intended to confine the scope of the inquiry to two important issues: (1) the payment by Ontario Hydro of $50,000 to John Aird, Jr., in connection with the purchase of the O'Brien Company's interests on the Mississippi and Madawaska Rivers; and (2) the purchase of the Dominion Power and Transmission Company for $21,000,000—a price greatly in excess of the figure at which the company had been willing to sell to Ontario Hydro many years before. Subsequently, at the insistence of Hepburn and the leaders of the other Opposition parties, the inquiry was extended to cover the payment by the Beauharnois Corporation of

$125,000 to Aird after a contract had been signed with Ontario Hydro for purchase of power, and the relation, if any, between these two payments to Aird.

Within these terms of reference the Royal Commission was appointed on August 19, 1932, and it presented its report on the following October 31.

What emerged in evidence was that Aird, having supplied much information regarding the properties in the course of various attempts to arrange a contract between Ontario Hydro and the O'Brien Company, stood aside when his attempts failed, and by arrangement, after other negotiators had concluded a contract, was paid $50,000 by Ontario Hydro for such services as he had rendered. R. O. Sweezey, President of the Beauharnois Power Corporation, paid Aird an additional $125,000—$120,000 in Victory Bonds and the remainder in cash—for, according to Sweezey, the Ontario Conservative Party. Aird, in evidence, flatly contradicted Sweezey, declared he had no association with any political party, had never intended that any of the money should go to political funds, and that the money had not been given him for that purpose, but only for professional services rendered under contract. There was no evidence that any political party had, in fact, benefited by any part of the payment. After careful scrutiny of all the circumstances, the Royal Commission reported that the Sweezey payment to Aird had no connection with the purchase of power by Ontario Hydro.

As for the purchase by Ontario Hydro of the Dominion Power and Transmission Company's assets, in this case too the Royal Commission unhesitatingly found that the purchase was in the public interest, and made after full and adequate investigation; that the price was reasonable and had not been prompted by any motive other than the public good. In fact, negotiations had been going on from 1917 until their conclusion in 1930. It was shown that the price was reasonable in relation to the benefits which possession of the property would bring to Ontario Hydro, including such advantages as peak power at lower cost than was possible by any other means; preventing the DeCew Falls GS from being acquired by hostile competitors of Ontario Hydro; avoiding duplication of plant and services; avoiding installations of equipment for 25-cycle energy for Ontario Hydro customers and 66⅔ cycle for the company's customers; preventing the owners from undertaking new construction which would have raised the price of eventual purchase; and the prevention of later competition for water-rights. The Commission was completely exonerated of any wrong-doing; on the contrary, the Commission's business dealings had been conducted on "the highest business principles and with great skill and rectitude." Ontario Hydro's engineers were also found to have carried out their part of the work with great professional competence.

Speakers at the Liberal Annual Meeting of November 18-19, 1932, again led by Hepburn, repeated the by then well established pattern of condemning Ontario Hydro's Quebec contracts, though on this occasion they steered clear of the issues on which the Royal Commission had reported. Since from a long-term point of view the necessity for importing large blocks of power to supplement that generated in Ontario Hydro's own systems could not be seriously doubted, it may be assumed that criticism of Ontario Hydro's basic purchasing policy was designed primarily to weaken the public's confidence in the Commission, preparatory to further verbal onslaughts.

Early in 1933, Opposition attention was deflected from a direct preoccupation with Hydro's affairs to consider the propriety of Premier Henry's participation in the negotiations for the purchase of the Ontario Power Service Corporation's Abitibi Canyon development. Henry had personally conducted the negotiations, successfully concluded in 1932, though he himself held bonds in the Corporation to the extent of $25,000 and an insurance company of which he was a director held another $200,000. He explained to the House that he had bought the bonds long before the question of government purchase arose, without access to any information other than that available to the public at large; having them in his possession when he came to conduct negotiations, he could not sell or dispose of them without risk of such action being misinterpreted in a way prejudicial to the value of the bonds.

Returning once more to a frontal attack, Opposition members, alleging improvident spending by Ontario Hydro, demanded in the House that the amounts paid to higher officials should be tabled, with a statement as to whether there had been reductions since the depression. J. R. Cooke, in his capacity both as Ontario Hydro Chairman and a Cabinet minister, refused the demand on the ground that a charge on the Consolidated Revenue Fund of the Province was not involved, and that therefore there was no reason why the information should be given in the House. The executive of the Ontario Municipal Electric Association, he concluded, was the only proper authority to receive it. There were friends of Hydro who felt that this reticence was, in the disturbed and suspicious political climate of the time, unfortunate from a public relations point of view. Cooke may have thought so too; in any case he did, in fact, give the information to the executive of the OMEA who, on a two-thirds vote, decided to make it public. On May 12, after the close of the session in the Legislature, yearly salaries of ranking Ontario Hydro officials were quoted in the Toronto *Mail and Empire* by T. J. Hannigan of Guelph, Secretary-treasurer of the OMEA.

In retrospect Ontario Hydro's "excessive" salaries do not seem particularly high, yet in the rancorous economic disillusionment of the early thirties, they seemed to many people excessive, and much more so by prevailing civil service standards. The Chairman received in 1933, $13,780 and the two other Commissioners $8,280 each. The chief engineer was paid $24,726 plus $5,000 for his services as manager of the Wanapitei development; the chief hydraulic engineer received, in 1932, $13,894; chief construction engineer $12,815; chief accountant $11,194; treasurer $9,395; secretary $8,944; and other executives from $5,225 to $11,194.

On August 31, 1933, Chairman Cooke issued a formal statement clarifying Ontario Hydro's position on power supplies and demand. The statement noted that while the total generated and purchased power for all systems was 1,760,052-hp, the total load-meeting capability was only 1,585,000-hp, against actual peak demands and contractual obligations of 1,289,267-hp, so that the special reserve for recovery from depression amounted to just under 296,000-hp. As industrial production was even then recovering at an unexpected rate, it appeared likely that, despite annual purchases of some 790,000-hp of energy from Quebec, by 1937 Ontario Hydro would have difficulty keeping pace with demand. In support of these observations Chief Engineer Gaby published in Toronto a work entitled, *Trends of Electrical Demand in Relation to Power Supplies*, a reprint of an address he gave to the AMEU and OMEA on January 25, 1933.

As the election campaign of 1934 approached its climax, so did the political storm that raged about the heads of Ontario Hydro's three Commissioners, Cooke, Maguire and Meighen. Hepburn, in final bursts of moral indignation, reiterated and compounded in successive speeches the various allegations of incompetence, mismanagement, and malfeasance that had been made formerly. Despite Chairman Cooke's responsible statement of August 1933, the Liberal leader accused Ontario Hydro of having surpluses of up to one and a half million horsepower. The Commission countered by issuing a pamphlet, *Paid-for Propaganda*, prepared in the chief engineer's office. The pamphlet warned the public against the anti-Hydro campaign still being waged by the National Electric Light Association and its supporters, but it also charged Liberal critics of Hydro with having deliberately adopted the misleading and inaccurate statements of the anti-public-ownership propagandists, thus, perhaps inadvertently, projecting the Commission and the engineering executive into a partisan political platform.

Shortly before the election, speaking at a banquet in Toronto, Henry suggested that the Liberal Party had received large sums of money from the Beauharnois Power Corporation for campaign purposes, and appealed for support of Hydro which, he stated, could not be maintained if subjected to political criticism and control. Hepburn in return again accused Henry and Meighen of an improper personal interest in the Abitibi purchase. He described it as a "swindle," claimed that Hydro was "fairly staggering from a mass of junk," and promised to give it a "courageous and economical administration."

On June 18, 1934, the voters of Ontario elected the first Liberal Government in twenty-nine years with Mitchell F. Hepburn as Premier. Less than a month later Ontario Hydro paid the penalty for its alleged involvement in politics. The Chairman, J. R. Cooke, and Commissioner C. Alfred Maguire were removed by order-in-council on July 11, Arthur Meighen having previously resigned on May 18. F. A. Gaby was retired after twenty-seven years' service, twenty-two of them as chief engineer. With him went I. B. Lucas, general solicitor; A. V, White, consulting engineer, and E. A. Hugill, head of the Right-of-Way Department; leaving Major Pope, the secretary, as sole member of the old guard to remain. Forty-three other Ontario Hydro officials in receipt of salaries of $5,000 a year or more had their contracts terminated, but were allowed to remain with the organization at such new salaries and conditions as a newly appointed Commission might decide.

Frederick Arthur Gaby, Chief Engineer from 1912 to 1934, with complete charge of engineering matters and reporting directly to the Commission, was in effect the architect of Ontario Hydro's immense power-plant developments and transmission systems. Like most other Ontario Hydro engineers, he was a graduate of the University of Toronto, having taken a degree in Mechanical and Electrical Engineering in 1903 and his B.A.Sc. in 1904. In 1907 he joined Ontario Hydro as Assistant Chief Engineer, becoming Chief Engineer five years later. He had worked in close collaboration with Adam Beck throughout Hydro's formative years, always ready with the statistics and technical data needed by the Chairman to counter and refute the arguments of those who opposed the public-ownership movement. In recognition of his outstanding achievements in the field of hydro-electric power engineering the University of Toronto awarded him in 1924 the degree of Doctor of Science. Shortly after his departure from Ontario Hydro he was appointed Executive Vice-President of British American Oil Company Limited, and he was a director of that company at the time of his death in 1947, in his seventieth year.

On July 12, 1934, the day following the wholesale dismissals, Premier Hepburn pressed home his attack by the appointment of a Royal Commission to inquire into the Abitibi purchase of 1932, a matter which had already been fully aired in the Legislative Assembly. The members of the Commission were Mr. Justice Latchford of the Ontario Supreme Court and Mr. Justice Smith, retired. The ex-Premier, George S. Henry, and the ex-Commissioner of Ontario Hydro, the Rt. Hon. Arthur Meighen, were both charged with conflict of interests in the negotiations of the purchase. As Premier, Henry had acted for the Government although, as previously stated, he personally held $25,000 of Abitibi bonds and was also a director of an insurance company which held $200,000. Meighen's position was substantially the same. He owned, personally, bonds in the amount of $3,000 and financial companies which he represented or owned held another $300,000 worth.

The Latchford-Smith Commission found that Henry "was precluded by these interests from taking part in the negotiations and resulting purchase." As for Meighen's involvement, the Justices concluded: "It was open to Mr. Meighen to have said to the other two Commissioners, the Government and the public, that he was interested personally and on behalf of his companies in bonds that the Government was requesting the Commission to purchase, and that therefore he could not take part in the recommendation or in the purchase. He failed to do this and was placed thus in the position as Commissioner of being a buyer of these bonds and being a seller of them in his individual capacity and as a director or manager of the companies in which he was interested."

Between July and December, 1934, Premier Hepburn appointed no less than seven Royal Commissions to inquire into Conservative iniquities. Besides Hydro's Abitibi acquisition they were: the Temiskaming and Northern Ontario Railroad; the Niagara Falls Park Commission; the Provincial Air Services; the Liquor Control Board; balloting in Toronto's St. Patrick's Riding; and the Ontario Athletic Association. It would appear, therefore, that his blitz of Hydro was directed less by personal animus than by broad strategic policy. As if to give the lie to so gentle an interpretation, he affirmed his political intentions by firing the Government's representatives on the Toronto and Ottawa Hydro Commissions: in Ottawa, newspaper publisher P. D. Ross was replaced by the President of the East Ottawa Liberal Association, Dr. Rufus H. Parat; and in Toronto, Loftus H. Reid, Chairman of Toronto Hydro, was replaced by Kenneth A. Christie, a lawyer who had worked for the Liberal Party, and who, it can be said, proved himself a painstaking administrator of the local system. The Toronto

Hydro Chairman refused to yield without a fight. He had been re-appointed in April 1933 for a two-year term, and he appealed his summary dismissal to the Ontario Supreme Court. Mr. Justice McEvoy found in his favor with costs, meaning that Christie was guilty of usurpation. Premier Hepburn's riposte was to have Christie file an appeal, which kept the issue open until 1935, when a retroactive amendment to the Power Act was passed, making his appointment legal. Twelve years later, three years after the Conservative Party under Premier George Drew was returned to power, Loftus Reid was restored as Chairman of Toronto Hydro.

The Hydro-Electric Power Commission of Ontario, as reconstituted by Premier Hepburn, consisted of T. Stewart Lyon, Chairman, with the Hon. Arthur Roebuck and the Hon. Thomas B. McQuesten as fellow-Commissioners. A. Murray McCrimmon was appointed later as Controller; T. H. Hogg became Chief Engineer, Hydraulic and Operation; and R. T. Jeffery became Chief Engineer, Municipal Relations and Rural Power.

Lyon, who resigned his position as editor of the Toronto *Globe* to become Chairman of the Commission, had been one of Adam Beck's most steadfast and valued supporters during Hydro's early fighting years. According to John R. Robinson, editor of the Toronto *Telegram*, Lyon's advocacy of Hydro cost his paper "hundreds of readers and thousands of dollars of advertising revenue." Beck had frequently visited the *Globe*'s office to discuss Commission policies, whereas Beck's successors had stayed away, Lyon finding himself in frequent disagreement with their views. By his appointment to the chairmanship of the Commission, Lyon was enabled—and probably instructed by the Hepburn Government—to undo much of what Beck's successors had done. He and his fellow-Commissioners became the instruments of Hepburn's drastic program of purges and contract cancellations. Both Roebuck and McQuesten could be relied upon to support this program. Roebuck had been one of Hydro's public critics, and he was appointed Attorney General in Hepburn's Cabinet, while McQuesten, the third Commissioner, became Hepburn's Minister of Public Works and served also as Chairman of the Queen Victoria Niagara Falls Park Commission, also reconstituted by the Hepburn regime.

Both Roebuck and McQuesten served in Ontario Hydro without additional salary. R. T. Jeffery, the new Chief Engineer, Municipal Relations and Rural Power, had joined Ontario Hydro in 1913 as assistant municipal engineer of the Niagara System. A. Murray McCrimmon did not receive his appointment as Controller until mid-November. He was a utilities executive, a native of Kincardine, Ontario, who had worked in the traffic department of the Brazilian telephone service, and had returned to Toronto in 1928 to act in association with a stock exchange firm of brokers as consultant on bond securities. He came to Ontario Hydro soon after the Power Commission Act was amended to authorize Ontario Hydro to issue its own bonds. His function as Controller, as a sort of operating critic, included the duty of checking all estimates and budgets of expenditures as well as supervising of equipment, materials and supplies.

The government policy was to turn from a heavy-spending, "construction" era to a commercial or "new sales" phase. One of the first retrenchment efforts was the order to modify the plans for the projected new seventeen-storey Head Office in Toronto, on which construction started that year. Building was allowed to proceed, but to a height of only six storeys and at a cost of $1,150,000 instead of $1,900,000 as originally planned.

Commissioner Lyon also announced some "retrenchment" with respect to Ontario Hydro's past involvement in politics. The newspapers reported the discovery, in Ontario Hydro's files, of detective agency reports, paid for by the previous Commission, which had tried to determine whether or not money was contributed to the Liberal Party during the election campaign by private electric interests in the United States or in Canada, to assist the campaign against the Conservative Party and Hydro. Mr. Lyon announced that action was being entered against two former Commissioners, Meighen and Maguire, and against Gaby and Lucas, to recover $4,300 expended on authority of the Commission for the detective reports that had been found in the files.

Apart from the retributive dismissal of Gaby and other Ontario Hydro engineers, the Hepburn regime appears to have interfered little with the Commission's internal affairs. It is perhaps only coincidence that in 1934 an attempt was made to abolish Ontario Hydro's Construction Department and have the work carried on entirely through contracts with private firms. The Department's staff was reduced and to some extent redistributed. A department was organized for the disposal of material which had been bought and stored for use in planned future developments, but which now was to be sold off. This may have been part of the policy to end the "construction era." In any case, T. H. Hogg, the new Chief Engineer, was opposed to it, and it was largely through his efforts that the Construction Department was not abolished at the time.

There is no evidence that the professional standards and integrity of Ontario Hydro's technical staff were compromised in any way by these abrupt changes in personnel. Hogg, in Gaby's place, was another of Ontario Hydro's— and the world's—great engineers. He was born at Chippawa and had seen the beginning of hydro-electric development at Niagara. After he joined the Commission in 1913 as assistant hydraulic engineer he became one of the brilliant, hard-working team that included Harry Acres and E. J. T. Brandon. Together they helped to solve Hydro's multitude of highly complex novel technical problems. It was predictable that the tradition established by these men would not suffer under Hogg's chieftancy.

Another offshoot of the depression, one indicating the rapidly changing—and improving—status of Labor in industry, was the formation of a collective-bargaining unit among the employees of Ontario Hydro. Before 1935, collective bargaining relations between the Commission and groups of its employees were the exception rather than the rule. In that year, the Commission accepted the suggestion of a group of its Toronto employees to establish some sort of machinery for collective representation of the employee body, and the Employee Representation Plan was jointly developed. While the original Plan bore little resemblance to any of the current collective agreements between the Commission and its bargaining groups, revisions of the plan were made in 1945 to conform with the newly enacted Labor Relations Act. Shortly after the revision to the Plan, the Employees' Association evolved and a collective agreement between the Commission and the Employees' Association was signed in 1946.

REPUDIATION AND
RETRENCHMENT

23

With Premier Hepburn having to make good his election promises, it was inevitable that the contentious contracts between Ontario Hydro and its Quebec suppliers should have come up for revision.

Unquestionably changing circumstances had made a rewriting of the contracts advantageous to Ontario Hydro. A passage in the 1935 Annual Report reads: "At no time since the contracts of 1929-30 were made has there been need to call for more power from the Quebec contractors than could have been supplied by the Gatineau Power Company. For this unused and unusable power the Commission had been paying an annually increasing power bill which for the year 1929 totalled $1,639,516, and which has increased, as at October 31, 1935, to $8,232,968. Had the contracts remained in force the payment to the four Quebec contractors in the year beginning November 1, 1936, would have totalled $10,965,000 on a tendered contract supply of 731,000-hp." Chairman Lyon maintained that because of these contracts, withdrawals were made from the contingencies reserve in each year from 1932 until and including 1935.

To cancel the contracts, legislation was necessary. The contracts involved were as follows:

(a) Gatineau, five contracts dated May 19, 1926
    Gatineau, one contract dated July 27, 1926
(b) Gatineau, two contracts dated December 28, 1927
(c) Beauharnois, one contract dated November 29, 1929
(d) Chats Falls Power Company (Ottawa Valley Power Company) one contract dated February 15, 1930 and one contract dated February 24, 1931, known respectively as the "Power Contract" and the "Operating Contract."
(e) James Maclaren Co. Ltd., one contract dated December 20, 1930

210

James Maclaren Co. Ltd., one contract dated January 14, 1931

Early in the year, Attorney General Roebuck prepared the ground in a series of radio broadcasts for what was to be the most contentious measure of the legislative session: the 1935 Power Commission Act. Ontario Hydro, according to the Attorney General, had bought huge amounts of electric power for which there was no existing use nor any probable use for many years. In the meantime, Ontario Hydro had saddled itself and the municipalities with ruinous expenditure, and contracts believed by the Government to be illegal, void, and unenforceable—this being Roebuck's re-statement of what the Liberals had been saying while they were in Opposition. In 1926 and afterward, the Attorney General intimated, Ontario Hydro and the corporations, without the consent of municipalities or the ratepayers, and contrary to the rights of the municipalities under the Power Commission Act, had entered into contracts for long periods, for large quantities of power generated outside the province, and regardless of whether it could be used or was desired. In this way, Ontario Hydro had become liable for large payments which, when charged to the municipalities, had raised costs illegally.

Premier Hepburn introduced the bill, which was given its first reading on April 1. It met with vigorous opposition. He moved the second reading a week later on April 8, and made a fifteen-minute speech declaring that the Government was determined on its course, that the public regarded this attitude as a great, courageous stand, that Ontario credit was not endangered, and that there were ample and most respectable precedents for such action. He instanced the British repudiation of war debts to the United States, but without mentioning Germany's prior reparations failure. He instanced the Ontario Government's 1916 cancellation of the Electrical Development Company's franchise at Niagara Falls.

There ensued a 26-hour debate, one of the longest continuous debates in the history of the Ontario Legislature. On April 11, Mr. Hepburn moved the third reading and the bill was passed by a division— 57 for, 17 against. It became Chapter 53 of the 1935 Statutes, to be brought into force later by proclamation. The delay was designed to encourage and give time for the negotiation of revised contracts and also, initially, to protect the Eastern Ontario Division against power shortages pending completion of a frequency changer at Chats Falls GS. The Act declared the contracts with the Gatineau Power Company, Beauharnois Light, Heat and Power Company, Ottawa Valley (Chats Falls) Power Company, and James Maclaren Company, to be and always to have been illegal, void, and unenforceable as against the

Hydro-Electric Power Commission of Ontario, and that no action founded on any contract thus declared void, or arising out of performance or not of its terms, could lie against the Commission.

Passage of the Act occasioned an uproar in the financial community, although most of the larger bondholders affected had already anticipated the effects of the cancellations. One bank, on March 29 and 30 and April 1, just before the introduction of the bill, sold bonds to the value of $350,000 of the Beauharnois, Maclaren and Gatineau Companies. In the House, on the budget debate, demand was made for investigation of an alleged "leak" which allowed brokerage firms to unload the Quebec power bonds before an official announcement of Government policy. But the Ontario Securities Commission reported on April 16 that there had been no leak, that the financial people regarded the policy as "a well-kept secret." As the bankers explained, they had been privileged to hear the Attorney General's radio addresses.

The Canadian Bankers' Association advised the Government that it "viewed with alarm" the arbitrary cancellations which were bound to be damaging to the credit of Ontario and of Canada. Prime Minister R. B. Bennett, in the Federal Parliament, said the same thing. So did the Conservative newspapers—and not only in Canada. In England the *Times* on May 13 deplored the Act as "a blow to Canadian credit." The *Morning Post* found such action "almost unthinkable." The *Daily Telegraph* and the *Financial Times* were similarly shocked and disapproving. The *Globe*, Toronto, took the matter lightly.

On June 11, the *Globe* reported that Ontario Hydro had ordered a stop to any increase of Quebec power imports dating from July 1. Immediately afterward the Provincial Government called for tenders for a bond issue of $15,000,000 of five, ten and fifteen-year bonds bearing respectively 2¼, 2½ and 3 per cent interest. There was no response, the bond houses explaining their lack of interest by saying that they had been put in a difficult position through the threatened repudiation of the power contracts. But businessmen and bankers asserted simply that the interest was too low for the bonds to sell. A long controversy followed. Premier Hepburn was irritated. "The plain issue," he stated, "is whether the country is to be governed by elected representatives or by the dictators in control of the machinery of money." Shortly afterward his Government succeeded in floating a loan of $20,000,000 direct to the public, but at a higher rate of interest.

On October 21 and 22 Ontario Hydro ceased to take power from the Beauharnois and Maclaren Companies, and from the Ottawa Valley Power Company at Chat Falls. "This," said the Annual Report, "involved an increase in the delivery of power from Niagara Falls to Toronto and Hamilton and a rearrangement of existing lines in the vicinity of Niagara Falls and Dundas, and of the system's protective relays."

The return of better times and the steadily mounting demand for power made revision and renewal of the contracts inevitable. Meetings of bondholders in Toronto and Montreal attempted without success to find an acceptable basis for renewal. To give ample time for negotiations "for power actually needed and no more," Premier Hepburn still withheld enforcement of the cancellation Act. Ontario Hydro rushed to completion the installation of a frequency changer at its Chats Falls GS, and made other dispositions. It then declared independence of Quebec power and asked the Government to proclaim the Act. Discussion still boggled over terms. Finally, on December 6, the Government put the Act into force by order-in-council and proclamation, with Hepburn declaring impatiently, "We're through. But the four companies can return to us if they like. They can go to Hydro if they so desire."

Later in December Ontario Hydro announced the placing of three new contracts, two with the Gatineau Power Company and one with James Maclaren Power Company Limited. The first of the Gatineau contracts called for a minimum of 100,000-hp at $12.50, and for 33,000-hp to be provided for immediate standby at $10 per hp per year, with the difference between these and 260,000-hp to be reserved for Ontario Hydro for $1.75 per hp per year. The Maclaren contract was for 40,000-hp at $12.50, and the second agreement with Gatineau was for 60,000-hp of 60-cycle power. These new contracts, besides fixing lower basic rates, eliminated certain bad features of the old ones, such as payment in United States funds, liability to increased taxation by the Province of Quebec, and limitation of export by Quebec. Ontario Hydro said that the new payments in fiscal year 1936 would be only $2.8 millions as against $9.5 millions under the original contracts. But the Cabinet objected and withheld consent because the cost of power held in reserve would, they said, raise the price of used power to $14.50. Negotiations were resumed, and on February 8, 1936, two Canadian contracts satisfactory to all parties were announced by the Chairman, T. Stewart Lyon. There was also a new contract with the Maclaren Company. All contracts were at the lower price, $12.50, and were subject to cancellation on written notice after ten years rather than on a long-term basis. The Gatineau contract was a variable demand agreement covering from 100,000-hp to 260,000-hp. The Maclaren contract provided for 40,000-hp instead of 125,000-hp as formerly. Both were for 25-cycle power. The 60-cycle power contract with Gatineau was rearranged so as to provide for a minimum quantity of 42,000-hp up to a maximum of 60,000-hp of firm power. Authority for these transactions was given by the Power Contracts Validation Act of 1936.

Negotiations were also opened for new contracts with the Ottawa Valley Power Company and the Beauharnois Company, but these failed when both companies appealed to the courts, claiming that the 1935 Act, empowering cancellation, was *ultra vires* the Ontario Legislature. In addition both companies claimed payment of large sums, $115,000 and $573,000 respectively, as due them by Ontario Hydro for power made available. The Ottawa Valley Power Company's action was dismissed with costs, but the judgment was appealed. Pending the result of this appeal the court reserved judgment in the Beauharnois case. By a majority the appeal court reversed the lower court's decision. On hearing this, Hepburn, on November 23, 1936, said, "This is a temporary hollow verdict. This government in the public interest will fight to a finish, and, I repeat, will never pay." He threatened appeal to the Privy Council, and further legislation.

In the end, agreement was reached. The new legislation in 1937 confirmed a new contract between Ontario Hydro and the Ottawa Valley Power Company, and amended the Joint Development Agreement of Chats Falls and the Power Contract and Operating Contract. The revised price was again $12.50 instead of $15 per hp per year. By 1938 a further Validation Act was required and passed to confirm contracts with the Gatineau Power Company for both 60-cycle and 25-cycle power made in December, 1937, and a contract of the same date made with the Beauharnois Company and one with the Maclaren Company.

All these negotiations were expedited by the continued upward trend of business from the middle of the decade to the outbreak of World War II in 1939. The revised contracts provided for long periods. The 1936 Gatineau contract for 60-cycle power was amended in 1938 to make available 60,000-hp and the contract for 25-cycle power was revised to provide 200,000-hp by November 1, 1938, and up to 260,000-hp by November 1, 1939, both contracts to expire November 30, 1970. The 1929 Beauharnois contract for 60-cycle power was amended to establish a minimum contract demand of 125,000-hp by December 14, 1937, increasing to 250,000-hp by November 1, 1943, at a price of $12.50 per hp per year, the agreement to run until November 1, 1976, and to be renewable by mutual agreement until the year 2003 A.D. The Maclaren-Quebec Hydro contract of 1930 and 1936, for 60-cycle power, was amended, also in 1938, to establish a minimum contract demand of 40,000-hp, to be increased by stages to a maximum contract demand of 100,000-hp by November 1, 1944; again the price was to be $12.50 instead of $15.00. This agreement was to remain in force until October 31, 1970.

The Commission, as reconstituted by Premier Hepburn in 1934, remained in office just long enough to accomplish the contract revisions for which he had campaigned. On November 1, 1937, Thomas H. Hogg, while continuing as Chief Engineer, became also Chairman of the Commission. The two newly appointed Commissioners were Hon. W. L. Houck, Vice-Chairman, and J. Albert Smith. R. T. Jeffery was retained, becoming Chief Municipal Engineer and replacing A. Murray McCrimmon as Acting Secretary and Controller.

The Vice-Chairman, William L. Houck, was a newly elected member of the Legislature, and had been made a Minister without Portfolio in Hepburn's Cabinet. J. Albert Smith was a member of the Liberal Party from Kitchener. This Commission—Hogg, Houck and Smith, with Osborne Mitchell as Secretary from 1938—was to continue without change in charge of Hydro affairs until the return of another Conservative Government with George Drew as Premier in 1943. Osborne Mitchell was an Englishman who had graduated in electrical engineering at London University and had come to Canada in 1923. He then joined Ontario Hydro, but resigned three years later to join Hugh C. Maclean Publications Limited of Toronto as editor of an electrical journal.

Renewed prosperity now justified Hydro's long-term plans to meet ever-increasing demands for power. In the last years of the decade, the mining and pulp-and-paper industries of the vast northern areas of the province were rapidly expanding. In the rural districts, lines were extended and more and more loans were given to farmers, all townships having passed the necessary by-laws so that five-year agreements were everywhere available. In the first year of the new Commission's regime there were 13,117 miles of rural delivery lines and 86,194 customers including approved customers and lines authorized or under construction. Over 2,300 miles of new line were approved—as much as the total for the five-year period, 1932-36.

Similar expansion went ahead in that region north and west of French River, from the Quebec boundary to the Manitoba boundary and Hudson Bay, where Ontario Hydro operated its hydro-electric installations, except for the Thunder Bay plants, for the Province of Ontario. In 1934 the financial summary for the Northern Ontario Properties had shown a deficit of $337,754 which was charged to the Province, subject to repayment out of any future surplus earnings. Such earnings were soon to become available. The records speak of "new gold-mining customers added." Rarely do these customers go off the books, as happened in the Espanola District in 1937, when the only customer abandoned all mining and milling operations. By the time war broke out in 1939 the load increase was so great that the Northern Ontario Properties then being supplied included thirty-eight mining developments, two cities, four towns, nine villages, hamlets and mining townsites, and three rural power districts.

Typical of pioneering enterprise in the north country was the single generating unit operating to capacity at Rat Rapids to supply mines in Central Patricia. This station was later linked up with the Ear Falls GS on the English River, developed as a source of power for the mines in the Red Lake Area. Passengers, mail, express and light freight were taken in by plane, but the freight was usually transported by water from Hudson, a distance of one hundred and fifty miles, with four portages.

This was the first project on which the Commission used planes for transport. Teams and tractors could only be employed in winter time on the long haul from the railhead at Savant Lake, but in the summer time a good deal of open water had to be crossed, and the heavy material and equipment was brought in by scows and boats. Even then, there were many long portages over which Indian paddlers had to carry both canoes and cargoes. Using a tump-line, a good man could carry 250 pounds on his back, while a very strong man might manage 300 pounds or more. At Rat Rapids was used what may have been the last York boat to see service in Ontario. This was the type of craft used by the Hudson's Bay Company for taking furs out to James Bay and Great Lakes ports and for bringing in supplies to the various posts in the Northern hinterlands. The York boat, propelled by either oars or paddles, had a prow at each end and was very heavily built. Eight men constituted the normal crew and they were required to carry the half-ton boat as well as its load across very rough portages.

Although the Commission did not take over control of the Abitibi District until 1933, it built in 1931 a high tension line nearly 200 miles southward to Copper Cliff near Sudbury to supply the International Nickel Company. Much of this line was built through virgin forest for more than 135 miles. There were no roads over which supplies could be brought to the construction crews, so the Ontario Hydro engineers had to complete the roads themselves, as well as cut the right-of-way for the transmission lines. All steel had to be brought to location in the winter time, when the snow provided a sufficiently smooth surface for heavy traffic over the stumps. On the building of this line between 800 and 900 men were employed and no less than 102 teams of horses. During the winter it was necessary to protect the animals as well as the men from the severe cold—temperatures averaging 20 to

30 degrees below zero, and on occasion going as low as 50 below. The horses, heavily blanketed, were sheltered by canvas roofing and screens which, erected in the dense bush, protected them from the weather.

Mining prospectors combed the area, staking claims, opening mines, and calling for power. In the Abitibi district there were new installations at Abitibi Canyon GS, at the Hunta Switching Station, and at Iroquois Falls TS and Kirkland Lake TS. In Ontario Hydro's annual report we read of "the completion and occupation of ten new operators' houses, a new retail store, a new staff house, a new school and a new hospital, which relieved to a great extent the lack of suitable living and service accommodation for the permanent staff. . . . Where considered advisable, school-houses, recreation rooms, stores, hospital and other buildings are included and provision is made for water supply and electric service, sewage disposal and fire protection."

Of special interest among Ontario Hydro's northern undertakings in these years was the construction of the Long Lac Diversion some fifty miles east of Lake Nipigon on the north shore of Lake Superior. There, near the Height of Land on the rugged 1,800 foot peneplain high above the largest of the Great Lakes, rivers and lesser lakes that normally flowed northward to the Albany River and thence to Hudson Bay were ponded by great earth dams and made to flow southward into the Aguasabon River and thence into Terrace Bay on Lake Superior. These diversions were designed to provide more power for the Ontario Hydro generating stations at Niagara and to facilitate the transporting of logs for pulp and paper companies. This and subsequent changes in the course of northern rivers aptly demonstrates the truth of the observation once made by the American conservationist, Fairborn Osborn: "Man has now become a geologic force."

Expansion also continued in the long-settled, more densely populated areas of Southern Ontario. There it was necessary to provide for the summer migration from the cities to the Lakes. In the account of the Georgian Bay System in 1938, the "resort" names abound: the South Muskoka River was in good flow, so that three plants on the river increased in output and this, in conjunction with the tie-line between the Eugenia and Severn Districts, meant that no difficulty was experienced in distributing 5 per cent more energy than in 1937. Summer peaks due to the tourist trade were met by buying power from the independent Orillia municipal utility. The Ragged Rapids GS was brought to completion at this time.

During the pre-war years there occurred a number of disasters, including the 1938 flood which crippled the Niagara System and caused much public apprehension. The ice jam and the height of water in the river far exceeded all previous records and created a serious threat to the Ontario Power GS, the destruction of which would mean that light and power would be cut off from many parts of Ontario. Hydro engineers were less alarmed. Admittedly, rehabilitation of the damaged generators was a major problem, but fortunately the Commission had sufficient capacity to allow for the transfer to other plants of the 180,000-hp load which had been supplied by fifteen disabled machines.

In the city offices of Ontario Hydro, as in the field, the stir of better times quickened the tempo. Part of the Electrical Engineering Department was established as a planning section to anticipate the increasing demands for power and to carry on the necessary overall planning of power developments to meet them. In the Laboratory and Approvals Department the volume of business increased greatly. The Electrical Inspection Department found its work so increased that it too had to re-open some offices which had been closed during the depression years. The five-year-old Research Committee became so busy that it had fourteen subcommittees working in conjunction with the Laboratories. The Canadian Electrical Code, Part One, released by the Canadian Engineering Standards Association, had been adopted by all the provinces, thereby assuring an equal standard for all wiring and equipment installations throughout Canada. The Code brought about improved and safer electrical construction of many types of equipment and the Department's label service was now used by factories in Quebec and Ontario for a great variety of equipment.

The Inspection Department had so grown in its twenty years of existence that it now employed a large staff of inspectors—in its sixteenth year, 1931, it already had sixty-three inspectors working in thirty-two districts. At that time they were much needed to control the sale of dangerous appliances, such as certain inferior types of water heaters and, again in 1936, the inspectors were kept busy following a change of tariff regulations which allowed many electrical appliances of unapproved type to be brought into Ontario. During the decade, many United States manufacturers of appliances established branch factories in Canada and applied for approval service, particularly for hairdressing equipment, refrigerators, oil burners, and wiring devices. The volume of work of the Department was largely determined by new building construction in the province. As the depression receded and new buildings began to go up once more, the Department found itself with all the work it could well handle.

In 1938 Ontario Hydro completed arrangements which enabled the Commission to report "ample supplies of power secured for some years ahead." In consequence, said the Report, "the Commission plans to enlarge its activities in the promotional field and has appointed a director of Sales Promotion who is organizing a department to be responsible for advertising and promotional work. This department will form a coördinating medium uniting the efforts of the municipal Hydro utilities, the HEPC, and other branches of the electrical industry in the Province. By such united effort the maximum results should be obtained."

Ontario Hydro's report for 1938 seemed badly dated soon after it came off the press in July 1939. In September, Hitler's bombers roared over Poland, and again Hydro and all its machines and men were harnessed to the service of war.

HYDRO MOBILIZES
ITS KILOWATTS

24

In September 1939 Hydro took prompt steps to mobilize its resources in support of the war effort. Shortly after, the Commission set forth its policy in a public statement:

In all departments first place has been given to the task of ensuring that for the war industries of Ontario there should be ample supplies of power available wherever and whenever needed.

This priority was strictly maintained throughout the war and accounts for practically everything that was done by the organization during the first five years of the decade. First concern was given to defensive precautions against possible enemy attack and especially against sabotage of vital electrical installations. These were surrounded by high chain-link fences; men who worked in the plants were made special constables; armed guards were on duty day and night; and passes were issued to persons having business in the plants and were strictly scrutinized before anyone could go in. Later, in Toronto and other large communities, there were practice air-raid warnings, with all the spectacular paraphernalia of sirens, searchlights, blackouts, and manning of emergency shelters and stations by volunteer wardens. In case of attack the local Hydro personnel were instructed to make instant disconnection of all street lighting.

The pinch of power shortage was soon felt. To lighten the peak overload, an order-in-council required municipalities which had adopted daylight-saving time to continue to use it in winter as well as in summer. Later, power restrictions were directed by a Power Controller appointed to the Federal Department of Munitions and Supply.

At the start of the war, Ontario Hydro's resources amounted to 1,558,500 kilowatts. Primary power needs were for 1,317,000 kilowatts—more than the Commission's total resources had been in 1918 when World War I ended. To provide additional power for essential

220

services three new developments were undertaken: Big Eddy on the Muskoka River; Barrett Chute on the Madawaska River immediately above Calabogie Lake, where two units of 28,000-hp were installed; and a new generating station adjoining the old plant at DeCew Falls.

With the collapse of France the pressure increased. At the turn of the year an additional 60,000-hp was obtained from the Gatineau Power Company, and in July 1940, 20,000 additional hp from the Maclaren-Quebec Power Company, with the date of delivery advanced from November. The St. Lawrence project was revived and there were joint meetings of American and Canadian technical advisers. Arrangements were made with the United States so that Canada could immediately use additional water, equal in amount to that to be added to the Great Lakes by two large-scale diversions from the Albany River watershed north of Lake Superior but normally draining into Hudson Bay. The Long Lac project, which had been completed in 1938, was the first of these diversions, and toward the end of 1940 work was commenced on the second, the Ogoki Diversion. When finished, the total diversion would be 5,000 c.f.s. In an effort to cope with the expanding demands, the Commission's own systems were interconnected with certain municipal and privately owned local systems.

All this involved an amount of engineering and administrative work without precedent in the history of Ontario Hydro. New lines were being built, such as the new 110,000-volt single circuit line on steel towers between St. Thomas and Windsor. In the east of the province a new 220,000-volt line was built to carry current from the Quebec boundary to Toronto and Hamilton. All available power from Ontario Hydro's own generating stations and all 60-cycle power provided by contract from the Gatineau company was required for the eastern area. In the Northern Ontario Properties, too, the load growth in 1940 was phenomenal. Surveys for new power sites were made on the French River at Five Mile Rapids. In the Sudbury district the flow of the Wanapitei and Sturgeon Rivers was under study with a view to increasing the efficiency of the existing installations. In the Patricia-St. Joseph district a third unit was placed in service at Ear Falls. This station and the Rat Rapids GS were operated in parallel to meet the increasing load and to serve one interconnected network of transmission lines.

Farmers were urged to use more electricity in order to release manpower for war services and to produce more food at lower cost. Ontario Hydro's Sales Promotion Department prepared special engineering reports to assist firms making war materials. Ontario Hydro operated a free engineering advisory service on lighting, motor power and heat, which was welcomed by many industrial organizations throughout the province. The Testing and Research Laboratory gave valuable assistance to the Department of Munitions and Supply and to the United Kingdom Technical Commission dealing with electrical affairs.

By 1940 the emphasis of the Sales Promotion Department had shifted to advising plants in the efficient use of electricity via power factor, motor loading and lighting surveys. In domestic services, conservation of energy was the aim. Under the "Bits and Pieces Program" of the Dominion Government, introduced in 1941, Ontario Hydro was producing various miscellaneous pieces and parts of war equipment. In the rural districts Ontario Hydro curtailed line construction, and war financing ended the granting of loans for farm installations, while more efficient use of existing installations was urged.

The negotiations between Canada and the United States were concluded in 1941, allowing Ontario Hydro 14,000 c.f.s. of additional water diversion on the Niagara River, including 5,000 c.f.s. from the Ogoki and Long Lac Diversions. This augmented the flow available for power generation at hydro-electric plants on the Nipigon River, on the Niagara River and at DeCew Falls.

Studies of the St. Lawrence project continued. Agreements between Canada and the United States and between Canada and Ontario regarding the St. Lawrence development were signed, but they required legislative approval in the United States: since this was withheld, Ontario Hydro at once had to think of alternative developments. In September work began on the construction of the new plant at the DeCew Falls GS; the initial installation was to be one 65,000-hp unit with design allowing a total installation of 200,000-hp, 25-cycle. This development was made possible by the additional diversion of water now permitted. In the Georgian Bay district the National Defense Industries at Nobel called for more and more primary power. The Hanover Frequency-Changer Station was kept in constant operation all through the year. A total transfer of 43,982,000 kwh was made from Niagara to this system to meet primary needs, but surplus at off-peak periods was fed back to Niagara, where it could be used. In October the first unit of 5,000-hp went into operation, and a second one in November, at Big Eddy GS near Bala. At the Lakehead large blocks of power were being used at Port Arthur and Fort William for munition plants and shipbuilding. In other parts of the Thunder Bay System the output of all generating stations had to be restricted to primary power until October, when increased water levels made energy available for electric steam-boiler operation.

During the first year of the war Ontario Hydro spent more than eleven million dollars on new capital construction; in 1941 seventeen million was spent. For direct war production the Commission was supplying a half million horsepower, or about 25 per cent of its total output.

On December 10, 1941, in an address delivered at Princeton University, Dr. Thomas H. Hogg, the Commission's Chairman and Chief Engineer, presented a review of Hydro's origins and achievements. After discussing the Hyde Park Declaration, in which the Prime Minister of Canada and the President of the United States agreed on measures for a coördinating program for defense production, Dr. Hogg gave a statistical picture of Hydro. The Commission then owned and operated forty-six hydro-electric generating stations, with an installed capacity of over a million and a half horsepower, two-thirds of which was produced by stations designed and constructed by the Commission. From private companies 865,000-hp was being bought, making a total capacity of over 2,400,000-hp. Dr. Hogg gave the demand for primary power in December 1941 as about 2,200,000-hp. This included the requirements of 886 municipalities and rural power districts, 152 industrial corporations, and the firm export to the United States at Niagara Falls. In addition, between 50,000 and 100,000-hp was being exported to help vital war industries across the border, making the territory thus served by the Commission equal in area to one and a half times the State of New York. Capital invested in Ontario Hydro plants and properties totalled 363 million dollars. Capital investment of member municipalities in local distributing systems amounted to 99 million dollars, making a total investment of 462 millions. Against this investment there were reserves and municipal surpluses aggregating 242 millions. Dr. Hogg marshalled this and other data in support of his general argument in behalf of public ownership of essential utilities.

In 1942 the three Ontario Hydro generating stations at Niagara operated at maximum twenty-four hour output throughout the year, made possible by the added water diversion of 14,000 c.f.s. All sources of power supply available to the Niagara System were used to the greatest possible extent. The full contract amount of 250,000-hp of Beauharnois power was delivered, and the Maclaren company supplied 182,250-hp, including a temporary supplement of 57,500 for the duration of the war. Yet over a few daily peak periods in the winter of 1941-1942 it was found necessary to interrupt delivery to war industries for brief periods. By September, non-essential use of power was curtailed in the Niagara and Eastern Ontario Systems. Restrictive measures ordered by the Power Controller included the reduction of street lighting, elimination of sign, store window and showcase lighting, and prohibition of the use of electric air-heaters in stores and offices. In the Northern Ontario Properties, power was saved through a curtailment of gold mining. On the farms and in rural districts generally the slight increase in demand would have been much greater but for federal restrictions and voluntary conservation, and but for the practical cessation of construction except for war industries.

The pace of new electrical construction, however, was steadily accelerated. Just above the Falls, work was begun on the remedial submerged weir across the Niagara. Coördination and interconnection of the power facilities of the three Southern Ontario systems went steadily forward. In the Georgian Bay district the transmission voltage was raised from 22,000 to 38,000 volts over most of the system. The Eugenia Falls GS and Trethewey Falls GS were rehabilated and automatic control equipment was installed at Big Eddy GS to allow remote control from Ragged Rapids GS. At Barrett Chute GS in the Eastern Ontario System both units were in service by August, thus adding 54,000-hp to the system. This was the first of a series of carefully correlated power developments planned for the Madawaska River. During the year the storage dam at Bark Lake was completed. In the north, work went ahead on the Ogoki Diversion. The cost of all this raised the total expenditures by the Commission since the outbreak of war to a total of more than fifty million dollars.

In 1942 and in February 1943 the Commission signed power contracts with the Steep Rock Iron Mines and the Ontario-Minnesota Pulp and Paper Company. In the Ontario legislature, acts were passed in both years confirming these agreements. A rich deposit of high-grade ore had been found at the bottom of Steep Rock Lake, in the rugged bush country four miles north of Atikokan, between the head of the Lakes and Winnipeg. The deposits extended to a depth of two thousand feet below bedrock in the bed of the lake, and were estimated to ensure production at the rate of from eight and one-half to ten million tons a year for a period of twenty-five years.

To work these deposits required prodigious feats of engineering, including the emptying of Steep Rock Lake. This necessitated the removal of 110-billion gallons of water against an annual inflow of 5 billion gallons. It was one of the biggest pumping projects ever undertaken in the history of engineering. The capacity of the pumping plant was greater than the combined capacities of all the plants used in supplying water to Canada's fourteen largest cities. About 7,000-hp was required to operate this gigantic plant. Ontario Hydro supplied it at great cost by extending its high-tension lines in 1943 from Port Arthur to the mines, a distance of 120 miles. The mining company had to build seven miles of 44,000-volt main feeder line with subsidiary tap-offs from the Moose Lake terminus of the Ontario Hydro line to the main plant and pumping station.

Pre-war estimates by competent engineers had given the time needed to complete the preliminary work as from three to six years. But ways were found to speed up work schedules, and in spite of difficult wartime conditions, roads to the construction sites were built in the early months of 1943. By December 15 an astonishing job had been done: the pumping forebay and tunnel had been excavated, and the seven steel barge hulls, each carrying 24-inch pumping units, had been fabricated, finally assembled and placed in position, the discharge pipes had been connected with the forebay and the first pumps had been put into operation.

The power supply to Steep Rock was made available, indirectly, by the 50 per cent increase in water available for generating power in the Nipigon River. Water was obtained from the Ogoki Diversion, which was officially opened in July and was designed for an average flow of 4,000 cubic feet per second. This diversion made possible additional power at developed and as yet undeveloped sites and justified installation of the fourth unit of 20,000-hp at Alexander GS. A re-diversion was completed in the Abitibi district. Lumber companies had turned water from Lake Dasserat out of its natural watershed and into the Ottawa River. Now it was turned back again. In the district, increased demand for power for nickel refining was offset by the decrease in gold mining. Number one unit at Abitibi Canyon GS was shipped to DeCew Falls GS No. 2. There the single-unit new plant of 65,000-hp was officially opened in October to develop more power than the nine units in the original plant. Until this opening, all the generating stations on the Niagara River had been working at full capacity for twenty-four hours a day throughout the year.

Meanwhile, since coöperation by the United States in the proposed St. Lawrence development was still withheld, the Commission had turned to the possibilities of sites on the Ottawa River. In January 1943 it entered into agreements with the Governments of Ontario and Quebec and with the Quebec Streams Commission which gave Ontario Hydro all necessary rights to develop power at La Cave, Des Joachims and Chenaux. In February agreements were given legislative sanction.

In the Legislature, as a result of a Provincial election in August, 1943, the Liberal Government was replaced by a Conservative Government with the Hon. George Drew as Premier. This involved a change in the personnel of the Commission. On August 24 William L. Houck was replaced by the Hon. George H. Challies, a former

Cabinet minister now restored to office by Drew. On the same date, J. Albert Smith also ceased to be a member of the Commission, and W. Ross Strike of Bowmanville was appointed on June 16, 1944. Strike had been Chairman of the local Public Utility Commission and for a long time had been active in the OMEA; in fact he came to Ontario Hydro as representative of the OMEA.

In 1943 the agitation for a flat rate for rural districts, which in 1920 had been the subject of an inquiry by the Lethbridge Committee, approached its objective. The House requested Ontario Hydro to examine the causes of difference in the cost of power to the municipalities and rural districts. The Commission reported that the basic cost of generation was the same for all, but that other costs varied with distance of transmission, load and the number of users sharing facilities. To reduce this difference, three proposals were made: amalgamation of Southern Ontario, assistance to small municipalities with higher wholesale costs, and a new rural rate structure that would reduce the retail cost to 97 per cent of the rural consumers with substantial reductions to customers in the more remote districts. This new rate structure was expected to stimulate rural electrification after the war.

On January 1, 1944, the revised rural rates were put into effect. All rural power districts were rearranged into three large new ones: Southern Ontario, Thunder Bay and Northern Ontario. The local rural power districts were retained as administrative units but amalgamated into one division for pooling of expenses and revenues. A uniform rate structure was set up, with common rates applicable to each class of rural service except industrial customers, and with a simplified classification of consumers. Legal authority for this step was contained in the Rural Power District Service Charge Amendment Act, 1944.

On March 15, 1944, the OMEA exemplified the coöperative principle of public ownership in a resolution recommending to the coöperating municipalities that they accept a voluntary levy of not more than five cents per horsepower on their municipal loads, to help reduce the price to those municipalities whose power costs were in excess of a proposed maximum of $39 per horsepower. The Commission accepted OMEA's general recommendation, but because of the large aggregate load it was able to make the needed reductions by means of a levy of only two cents per horsepower for the fiscal year 1944.

In the internal administration of Ontario Hydro, the personnel in various departments were regrouped at this time as follows: 1. Executive and Secretarial. 2. Accounting. 3. Treasury. 4. Engineering—Operations. 5. Engineering—Design and Construction. 6. Engineering—Municipal. 7. Sales Promotion.

In the Annual Report the letter of transmittal for the first time carried a section headed: *The Commission and its Employees.* It spoke of the aim of Hydro as having been from the beginning to build up a staff whose ideals would include a deep sense of responsibility to the public and loyalty to the organization, intelligent and industrious application to work, and a high standard of technical competence. In return the employees were given a fair remuneration, security and continuity of employment.

Another event of this year was the enactment of authority to create The Ontario Hydro-Electric Advisory Council of five members, to be appointed by the Lieutenant Governor-in-Council. The appointments were not made, however, until July 1951, at which time the council was enlarged to nine members by legislative enactment.

In 1944 the Commission's operational structure was strengthened by the creation in Toronto of the office of Power Supervisor, for the purpose of coördinating demand and supply in the Southern Ontario System. The advantages of interconnection of systems were being utilized, and interconnection involved regulation of the amount of power generated in each division to provide for divisional fluctuations in demand and to regulate the transfer of blocks of power back and forth between divisions. Supervision of this process from Toronto was possible only through an intricate system of communications and supervisory equipment, and this was now devised.

The margin between available supplies and actual demand was very small. Constant vigilance and careful coördination of sources were necessary. The fact was that over the past ten years the Commission's business had practically doubled, its revenues from the supply of power and other operations had increased from over thirty million dollars to over fifty-eight million dollars, and the energy controlled by the Commission had increased from six and a half billion to more than twelve billion kilowatt hours. Various measures were being taken all the time to keep the demand-supply ratio in balance. In May a temporary additional diversion of 4,000 c.f.s. of water was made available for power production on the Niagara River and at DeCew Falls, and a remedial weir at Niagara was completed. In this System, the operations control at Chats Falls controlled Chats Falls GS, Barrett Chute GS, and plants of the Gatineau Power Company supplying the Commission. The output of these stations was regulated by specially designed equipment, maintaining control through carrier communication channels. As for the operations in the north country, there was in this year a general falling off in demand as the gold-mining industry continued to be curtailed. In the distant Patricia district there had been a 50 per cent decline since the beginning of the war. The staff at Rat Rapids GS had been reduced and the generating station operated only occasionally.

Half the decade had already gone by in a world at war. If peace should come it was inevitable that there must be a tremendous upsurge of new industrial activities, and there would be comparably acute shortages of power. Ontario Hydro engineers applied themselves to study all sources of possible load growth for ten or fifteen years to come, together with the transmission and other facilities needed to meet the foreseen demand.

THE POST-WAR SCRAMBLE
FOR POWER

25

In April of 1945 Hitler died and was buried beneath the ruins of his shattered Third Reich which was to have lasted a thousand years. The war in Europe was over, and four months later, after the dreadful light that shone over Hiroshima, it was over in the east. For the first time since 1939 the peoples of the western alliance could lift their eyes from the tasks of destruction and build again for the future.

In Toronto the Commission issued its annual report, foreshadowing what peace would mean for Hydro. "The lifting of war restrictions," read the report, "the return to standard time for the winter months, the ending of the need for rigid economy in the use of electricity in the home have, at least for the time being, more than offset the reduction in the demand for war industry." The total demand for primary power in the Southern Ontario System after VE Day was, in fact, greater than that of the corresponding months in 1944, while in 1945 the total energy output handled by the Commission was 65 per cent greater than it had been in 1938. The year, however, was marked by planning rather than by active construction. The Commission was still following what its Chairman, Dr. Hogg, had described as Ontario Hydro's wartime policy: "As the war progressed the Commission, by construction and purchase, augmented the power supplies to keep pace with demand, trying always to limit new construction to a minimum, in conformity with the national policy of husbanding the limited supplies of men and materials."

That cramping policy was soon to be discarded. The Commission had plans for eight major developments which were either totally new stations or extensions to existing stations. In the Southern Ontario System construction was started on a second unit, DeCew Falls GS No. 2: capacity, 70,000-hp, estimated construction cost over two years

229

$7,700,000. Construction was also started on a new plant at Stewartville on the Madawaska River, a plant similar to Barrett Chute GS, with a capacity of 54,000-hp costing $9,000,000. The third and most important project now authorized was the Des Joachims (pronounced D'Swisha) development on the Ottawa River about forty-five miles west from Pembroke. The initial construction was to include six units with a total capacity of 360,000-hp under a head of 135 feet to be created by a dam across the river, which would back up water to Mattawa, 57 miles upriver. Work was planned to begin in 1946, and the cost was estimated at $51,000,000. Provision was made for subsequent enlargement of the project to eight units. Most of these projects involved extensive transmission line construction.

In the Thunder Bay System a new development, made possible by the Long Lac Diversion, was authorized near the mouth of the Aguasabon River about fifty-three miles east of the mouth of the Nipigon. On the latter river at Alexander GS, a fourth unit, made possible by the Ogoki Diversion, was put into operation. As for the Northern Ontario Properties, in the Temiskaming district, in March, 1945, the Commission commenced to operate the plants of the Northern Ontario Power Company, bought on behalf of the Province for $12,500,000. These plants included eight hydro-electric stations with an installed capacity of 66,840-hp, 739 miles of transmission lines, 157 miles of distribution lines and 421 miles of telephone lines. The company was an amalgamation of the Cobalt Power Company and other interests which had developed water-power resources in the Porcupine district and on the Montreal River near Cobalt, when silver was discovered there. Five years later, in 1950, supplies of power to Northeastern Ontario were increased by the construction of the George W. Rayner GS on the Mississagi River near Thessalon, and by interchange facilities with the Southern Ontario System via Crystal Falls GS near North Bay.

A notable accomplishment of 1945 was the rate reduction to all gold mines for the first 5,000 horsepower, from as high as $45 in some cases to $27.50 per horsepower per year, a concession calculated to help the industry so badly hit by the war. In the rural districts throughout the province, where restrictions on new materials needed for construction were lifted in June, and where the Commission had plans for erecting about two thousand miles of new line—there were already 21,569 miles of rural lines serving 156,650 customers—power rates for industrial users in these areas were reduced wherever possible, effective January 1, 1946.

A reduction in 1944 to 4 cents per kwh to rural customers, due to the adoption of the uniform rate structure, proved so satisfactory in terms of increased demand that a further reduction was made in 1945 to 3.5 cents per kwh for the first block of energy. In the contract with municipalities, the formation of the maximum reserve fund by voluntary levy had enabled the Commission to cut the wholesale price of power to certain municipalities where actual costs were uneconomically high, and the consequent rate reductions led here too to increased use, larger revenues and lower unit cost, making further reductions possible. Of the sixty-three small municipalities estimated to be in need of financial assistance in 1943, and on whose behalf the introduction of a voluntary levy was proposed by the OMEA, by 1945 only twenty-one were left with an actual cost of over $39.00 per horsepower, and the original levy of two cents was again sufficient, as it had been the previous year.

The downward adjustment of rates approved by Ontario Hydro for domestic, commercial and power customers of 255 municipal utilities would benefit those customers by $3,300,000 a year. (The average cost to domestic users in 300 urban hydro systems was now only 1.15 cents per kwh.) Further, the Commission recommended and approved refunds to consumers in urban centres where surpluses had accumulated beyond what was required for new construction and maintenance. The surpluses and special reserves accumulated during the war were now to be used for deferred maintenance work and new construction. Municipal electrical utilities planned expenditures for these purposes in the immediate future amounting to $6,000,000. To assist in all this reconversion and reconstruction of industry, Ontario Hydro's publicity and promotional services were brought into action.

Another Ontario Hydro department called to assist in the work was the Research and Testing Department. About one-third of the staff were engaged in defense research for the Dominion Government at the end of the war. Their attention was now turned to problems of grounding, insulation, electronics, stress analysis, treatment of wood poles, and so on. Tests which yielded results of great operational and research value to the Commission were made in conjunction with the Central New York Power Corporation, via an existing 60-cycle, 110,000-volt line between Ottawa and Massena, New York, which was used to interconnect Ontario Hydro's Southern Ontario System with power systems in nine northeastern states and part of Quebec.

The general slowing down of the wartime armaments industry brought about by the cessation of hostilities could well have resulted in considerable surpluses of electricity over total demand, but any anxiety that may have been felt on this account was rapidly dissipated. As the 1946 Annual Report points out: "The well-planned reconversion of industrial plants to a peacetime program resulted in the power dropped by war loads being quickly re-employed." In addition, Hydro had to meet the requirements of an economy pressed into all kinds of new industry by the forced-draft acceleration of war, and because of this and the rapid growth of population, developing at fantastic speed.

Hydro's construction program was conditioned by these considerations. It called for generation and more generation, more and more transformer and distributing facilities. In the Southern Ontario System the chemical industry, electro-smelting and furnaces, and the cement industry were working to near-capacity, exceeding even their wartime production. New industries were manufacturing products never before attempted in the province. Domestic, commercial, industrial and municipal uses of electricity—all were increasing and leaving no reserve of unused generating capacity.

In the rural and northern mining areas the story was the same. Rural peak loads were 26 per cent higher than in 1945. The total number of consumers already served in the rural operating areas—in which the Government, by grants-in-aid, and Ontario Hydro had made a capital investment of $49,000,000—exceeded the number of Hydro consumers in all the cities of Ontario except the four with populations of over 100,000. In the Northern Ontario Properties, gold mining was being handicapped by shortage of labor, while in the Patricia district there was an increase of almost 30 per cent in the primary peak load, placing a strain on supply. At some places power was being purchased or brought in by frequency changers, as in the Temiskaming district where shortage was made up via Abitibi Canyon GS through a frequency changer set up at Kirkland Lake TS.

Forgotten was all the clamor about "improvident power accumulation" with which the politicians had harassed Hydro during the middle thirties. By October 1946 power demands had reached an all-time high. Shortage had become so acute that the Commission was forced to make cuts in delivery to customers purchasing power on an interruptible basis. It informed the municipalities that supply was insufficient and deliveries to basic industries were being seriously interrupted. The customers, urged the Commission, should be asked to coöperate voluntarily in reducing all unnecessary uses of current. Such appeals were made through the press and otherwise, but were disregarded by many users, although some relief ensued.

The new developments at DeCew, Stewartville and Des Joachims were making progress. At Des Joachims the preliminary work was giving the engineers some formidable problems. "The raising of the water level of the Ottawa River to elevation 500," says the Annual Report, "will make it necessary to clear about 11,000 acres of land between the site and the village of Mattawa, a distance of 57 miles upstream, and to relocate or raise 22.5 miles of railway trackages and 12 miles of provincial highway. Major items of construction include estimated quantities of 2,600,000 yards of material to be excavated, of which 90 per cent is rock, and 800,000 cubic yards of concrete." In the distant Patricia district work had begun on the fourth and last unit at Ear Falls GS, with a capacity of 7,500-hp, making a total installed capacity for the plant of 25,000-hp. At Rat Rapids GS, work had started on replacing a unit destroyed by fire in 1941.

A change in the constitution of the Commission was made during this year, through an amendment to the Power Commission Act which gave the Lieutenant-Governor authority to appoint a Vice-Chairman of the Commission. It also made some changes in the law pertaining to municipal utilities, stipulating that membership of local commissions in cities of 60,000 or over be made up of the mayor, one member to be appointed by the city and one by Ontario Hydro for a term of two years.

Another development that year was the formation of the Electric Service League of Ontario, composed of representatives of Ontario Hydro, local utilities, electric manufacturers, distributors, contractors and dealers. The chief object of the League was to foster a program of "adequate wiring."

The membership of the Commission changed. On February 28, 1947, the Chairman, Dr. Hogg, retired, but was retained as a consultant. His fellow-Commissioners, the Hon. George H. Challies and W. Ross Strike, remained as Acting Chairman and Vice-Chairman.

Thomas Henry Hogg was born at Chippawa, Ontario, on April 20, 1884, almost within sight of the Niagara River whose water power potential he was destined to have a leading role in developing. He graduated from the University of Toronto in 1907 with a B.A.SC. and received the degree of Civil Engineer in 1912. He served with the Ontario Power Company until 1911 and then for eighteen months was editor of an engineering journal before joining Ontario Hydro in 1913 as an assistant hydraulic engineer. In 1924 he became Chief Hydraulic Engineer and in 1937 was appointed Chairman and Chief Engineer, serving in these capacities until 1947. Dr. Hogg was awarded the degree of Doctor of Engineering, *honoris causa*, in 1927; in 1958, shortly before his sudden death at the age of 73, he was honored at the annual meeting of the Association of the Professional Engineers of Ontario by the presentation of the Professional Engineers' Medal for his "engineering skills, particularly in the development of electrical energy from water power resources," which had "contributed in rich measure to the industrial and economic growth of Canada and to the resulting benefit of its people."

In 1947 the load was the heaviest in Hydro's history and power shortage became more acute as the load continued to increase. Even after some 4.8 billion kwh had been purchased, cuts were needed, and voluntary cuts were not going to be enough. The municipal loads were such that in October the Commission found it necessary to obtain from the Legislature mandatory powers "to prohibit use of energy for specific purposes," and in November a modified system of power rationing for large industrial customers was introduced. Cuts toward the end of the year exceeded those which had been necessary in 1946. In the rural districts the pattern was the same, the peak load being three times heavier than before the war.

In the struggle to cope with this situation, substantial progress was made with the "greatest power development in Hydro history," a five-year plan designed to add within a few years a capacity of 943,000-hp, making a total available capacity of 3,691,000-hp. The total estimated cost of engineering projects approved by the Commission approached $320,000,000. A power purchase contract with the Polymer Corporation of Sarnia provided 26,000-hp. This was a plant built under Government auspices during the war to supply synthetic rubber when the natural supply was cut off by the Japanese—an emergency job of critical importance to the war effort, for which Ontario Hydro had loaned Richard L. Hearn, then Executive Assistant to the Chairman. The Niagara, Eastern Ontario and Georgian Bay Divisions were now interconnected; camp construction and preliminary work was proceeding on new developments at Chenaux on the Ottawa River, at Pine Portage on the Nipigon River, and at various other places. In the Thunder Bay and Patricia districts and elsewhere construction was being pressed.

The expansion program necessitated a reorganization of Ontario Hydro departments, for which consultant service was employed. The firm of J. D. Woods and Gordon Limited in Toronto was retained to make reports on administration and engineering. On Finance, G. T. Clarkson was retained to report not only on the financial aspects of frequency standardization, but on present practices in the determination of costs, and the financing of rural expansion and reserves.

A feature of the reorganization was the division of the province into nine regions, with regional offices in London, Hamilton, Niagara Falls, Toronto, Barrie, Belleville, North Bay, Ottawa and Port Arthur. Under the reorganized system the regional managers were "responsible for the operation and maintenance of all Commission properties and for all matters pertaining to consumer relations within their respective regions, and it is believed that the adoption of a decentralized type of administration will afford many advantages to the municipal and other customers." The officials of the regions were to consist of "a Regional Manager, Operations Engineer, Consumer Service Engineer, Accountant, Personnel Officer, and staff."

In 1948 Ontario Hydro was serving directly about two hundred industrial customers throughout the province with a monthly load of about a million horsepower. But the year was again marked by unprecedented demand, and the chronic shortage was now aggravated by a serious deficiency in rainfall. Water levels in the rivers and dams were abnormally low and cuts were made in deliveries from Quebec. As a consequence restrictions and a quota system were invoked. In Toronto, supply to the local system was curtailed by 350,000 kwh a day. By the end of March the restrictions were withdrawn because of lengthening daylight hours and the spring thaw, which made more water available. As fall approached, the public was asked by newspaper advertisements, radio broadcasts, letters to industrial users and, from the Mayor, to residential consumers, to limit use of electricity to the most essential needs. But rainfall was so far below normal that restrictions were imposed on September 14 and were made stringent by additional cuts almost week by week until the end of November, when late fall rains were heavy enough to ease the shortage and the restrictions.

Because another two years must pass before substantial alleviation of the shortage could come from major developments, the Commission decided meanwhile to install a number of relatively small steam-electric and diesel-electric units. In addition to emergency fuel-electric plants, which added 47,000 kilowatts of power to the Commission's resources, construction was begun on two major thermal-electric stations, the J. Clark Keith GS at Windsor and the Richard L. Hearn GS in Toronto. Their magnitude is suggested by the fact that Ontario Hydro's outlay on capital construction in the first three post-war years was equivalent to the combined total of like expenditure by the largest producers in the steel, chemical and oil industries of Ontario in the nine years between 1939 and 1948.

The five major hydro-electric developments on which work was being pressed were: Des Joachims, Chenaux, and La Cave on the Ottawa River; Pine Portage on the Nipigon River, and the George W. Rayner on the Mississagi River, imposing demands beyond the capacity of Ontario Hydro's own Construction Department. Referring to the problem, the Annual Report stated: "In a program of the present overall magnitude it was also found desirable to employ outside contracting firms to construct certain developments and important parts of others." Outside contractors were employed for the J. Clark Keith GS, the Hearn station, the Chenaux and George W. Rayner developments and the McConnell Lake Dam at Des Joachims. Besides the main development there and the project at La Cave (later named the Otto Holden), Hydro's own construction division built the two units of the Aguasabon GS, to be in operation by the end of the year, giving the Thunder Bay System a generating capacity in excess of 200,000-hp. Ontario Hydro also built the plant at Stewartville on the Madawaska where three units were in service by October, and the Pine Portage GS —the last available site on the Nipigon River.

On March 1, 1948, the chairmanship of the Commission, which had been vacated by Dr. Hogg twelve months earlier, was taken over by Robert Hood Saunders, CBE, KC, lawyer, gold-medal orator, sportsman and champion canoeist. In his profession and in public affairs Saunders had shown himself to be a man of great personal force and imagination. Before coming to Ontario Hydro he had been for many years a most active member of the Toronto City Council, having been elected and re-elected many times, as Controller, as Alderman, and for four terms as Mayor. In his years at the City Hall he helped to give Toronto and Canada their first subway and had worked energetically to promote low-cost housing and better relations with labor. In addition to an impressive roster of civic achievement, "Bob" Saunders had the common touch; in his manner, drive and indefatigable energy he bore a remarkable resemblance to Ontario Hydro's first Chairman. It was doubtless for this that Premier George Drew appointed him.

Richard L. Hearn became General Manager and Chief Engineer in 1947. He had been away from Ontario Hydro on various other engineering assignments from 1921 to 1942 and had been on loan from Ontario Hydro on two occasions subsequently—with the Polymer Corporation during the war and again in 1944 when he served as Canadian technical adviser to the Public Utilities Division of the Combined Production and Resources Board at Washington. In 1945 he was appointed Ontario Hydro's Chief Engineer of Design and Construction. A new secretary to the Commission was also appointed January 1, 1948, when Ernest B. Easson succeeded Osborne Mitchell, who had resigned to join the staff of the Brazilian Traction Light and Power Company. The new Secretary joined the Commission in September, 1930, after graduating in Commerce and Finance from the University of Toronto. He served Ontario Hydro first as Assistant Municipal Accountant, and in 1946 had been appointed an assistant to the Chief Engineer—Design and Construction.

Saunders and Hearn, together with the two overburdened Vice-Chairmen, Challies and Strike, made a formidable administrative group at a time when boldness of imagination and great tenacity and drive as well as sound judgment were essential. This team carried through a program of major construction; of emergency generating stations rapidly built and installed at strategic locations; of expediting, in addition to the major hydro-electric constructions, the thermal generating stations, the frequency standardization program in its earlier stages, and the re-activated negotiations over the St. Lawrence power project which had been the subject of discussion for nearly thirty years.

By 1949 work at Des Joachims GS and Chenaux GS was well in advance of the original schedule, while a new development, Sir Adam Beck-Niagara GS No. 2, was planned at Niagara. All told, the Construction Division now had three generating stations under construction at the same time—four more were being built by contractors—and was employing about 12,000 men, as compared with about 3,000 in 1919. A plan to form one great power pool by interconnecting the various systems was given substance in 1949 by the decision to build a tie-line from Des Joachims GS to North Bay in the Northeastern Region—there being as yet no physical connection between it and the Southern Ontario and Thunder Bay Systems and the Patricia district. Physical interconnection of the entire Ontario Hydro complex was now considered for completion over the next few years, making the Province a single integrated network.

Another notable event of the year was the use of helicopters for the inspection of long-distance transmission lines. Aerial survey and aerial photography had been introduced by Ontario Hydro in the mid-twenties in planning the location of the Abitibi and Gatineau lines; they were now used to patrol the transmission lines accessible only by foot, horseback or four-wheel drive vehicles. On June 15, 1949, Ontario Hydro became one of the first utilities in North America to use helicopters for line inspection. The saving in time and personnel, and the elimination of housing accommodation in remote areas, resulted in a reduction of some 25 per cent in patrol costs.

In 1949, also, a serious water shortage in Southern Ontario and Quebec brought back some of the wartime restrictions on power consumption. Consequently, restrictions on use for certain purposes were necessary in order to avoid excessive cuts to essential heavy industries. To offset the stringency, arrangements were made with the Niagara-Hudson Corporation of New York to waive for the greater part of each day the agreement for export of 45,000 kilowatts of firm power.

On the other hand, water conditions in the north were good, resulting in an increased output in the Thunder Bay System to a point 23.7 per cent higher than in 1948. New sources of power were provided by a second unit of 20,000 kilowatt capacity at Aguasabon GS, in service since December 15, 1948, and by the purchase on April 1, 1949, of the Kaministiquia Power Company's Kakabeka Falls GS, which had supplied power to the Commission for the Lakehead city of Port Arthur as early as December 1910.

In the rural areas the number of customers added in 1949 was 35 per cent greater than the 1948 increase. The average aggregate peak load of 228,800-horsepower was an increase of 18.4 per cent over that of 1948, meaning that the five-year post-war plan for rural expansion was actually exceeded in about four years. Paradoxically, Ontario Hydro was obliged to plan an upward adjustment of rates since revenues for the year from all rural customers did not cover costs. The raise was not confined to rural areas; in all systems the annual costs increased 11.4 per cent during 1949.

Most important to the scheme for interconnection with the power systems in Quebec and the states of New York and Michigan was the frequency standardization project—a plan to convert all electrical appliances in a 12,000 square mile area in Southern Ontario from 25-cycle to 60-cycle power. Originally Ontario Hydro's equipment was designed to take 25-cycle power, that being the frequency at which all but one of the first generating stations delivered power from the Niagara area in the earliest days of hydro-electric power development. Later, plants in the United States and elsewhere in Canada, and in many instances on Ontario Hydro's own network, were built for 60-cycle operation, which technical advances had rendered more desirable than the lower cycle power. As a result Ontario Hydro was constantly faced with the question of standardizing its entire system to a frequency of 60 cycles but successive proposals for such a relatively disruptive undertaking were shelved because of the complexity of the problems involved and the heavy expenditure it would entail.

In October 1945, however, at the initiative of Mayor Robert H. Saunders, Toronto Hydro asked Ontario Hydro what was being done about the frequency changeover, from 25 to 60 cycles, which had been discussed for thirty years. Could Toronto in the near future be supplied with 60-cycle current for the whole city, and at what cost? Ontario Hydro engineers studied the problem and in November 1946 made a preliminary report to the effect that there were "no insurmountable engineering difficulties involved" and that there would be "marked advantages in having standard frequency at 60 cycles." They estimated the cost of the changeover at more than $190,000,000 and suggested that there should be further consideration of technical and other problems by all concerned.

Their report had a mixed reception. At the request of the OMEA, the President of the AMEU appointed a committee of engineers and managers from the 25-cycle municipalities to discuss the problem.

The report of the AMEU committee, presented in February 1947, proved to be indecisive. The changeover, it said, would be too costly for all but some of the larger municipalities, would increase annual costs by 15 per cent and deplete reserves needed for other purposes. It recommended much more searching examination of the proposal. In March, at the annual meeting of the OMEA and AMEU, two men of great experience, J. Clark Keith of Windsor and E. V. Buchanan of London, read papers on the subject, the one in favor of conversion, the other against it. In April, in the Toronto *Telegram*, the former Chief Engineer of Ontario Hydro, Dr. F. A. Gaby, argued at great length against the desirability or economic feasibility of the scheme. Meanwhile the Commission continued its own studies with some outside help. On the engineering problem the Commission called in the firm of Stone and Webster Engineering Corporation; on the financial and economic aspects of the project, it turned to the Toronto firm of accountants, Clarkson, Gordon and Company. The reports of these consultants were submitted in December 1947 and January 1948. Both reports were favorable to the plan of executing the standardization program at that time. The OMEA, at a meeting in June, approved the scheme and the Commission decided to go ahead. It planned to start the actual work in 1949. Meanwhile legislative sanction was sought and in the 1948 session the House gave the Commission authority to transfer certain funds from its reserve account into a frequency standardization account.

On January 18, 1949, the work of changing all consumer 25-cycle appliances was started by Ontario Hydro and its contractors. The program, originally estimated to take fifteen years, was scheduled at implementation as a ten to twelve-year project—the greatest of its kind ever attempted anywhere in the world—at an estimated cost of $170,000,000 to Ontario Hydro and $20,600,000 to the municipalities. As the program became progressively enlarged, however, and the cost of labor and materials continued to spiral upward from the levels of the original estimates, the total cost to Ontario Hydro for the complete conversion operation was eventually $352,000,000. The latter figure provided for considerable expenditures not covered in the Clarkson, Gordon and Company report, including the costs of standardization in Northeastern Ontario.

Conversion was carried out in one municipality after another throughout the 1950's, and upon completion of the program in July 1959, over seven million appliances had been standardized for more than a million customers, including a small area in Northeastern Ontario.

INTERNATIONAL AGREEMENT

26

In his report to the Ontario Legislature on Hydro's progress during 1950, the Chairman of the Commission was able to make this statement: "The remarkable achievements which have been chronicled in the history of Hydro since its inception only forty-five years ago were completely eclipsed by the unparalleled accomplishments of the year 1950."

Robert Saunders had every reason for speaking on a note of triumph both on his own behalf and on that of the far-flung organization to which he had brought new energy and revived enthusiasm only two years earlier. At the beginning of his chairmanship Hydro was facing another of its power crises. The gigantic construction program launched in 1945 was behind schedule because of the post-war shortage of materials. Ontario was in the midst of the greatest population and industrial growth of its entire history. Yet two years later Saunders was able to point to such phenomenal progress that the perennial struggle for power seemed to be finally coming to an end.

During the preceding twelve-month period alone four new hydroelectric stations and five emergency thermal stations had been brought into operation, a more than substantial start had been made on the immense 60-cycle conversion project, and long-range planning had commenced for the interconnection of Ontario's separate systems and their eventual integration with those of Manitoba, Quebec, Michigan and New York to form a single mid-continental power grid. These accomplishments, however, did not stand alone. In the background had been, as always, the continuing hopes for the development of the St. Lawrence Power Project. Changing American opinion had brought realization of the century-old dream nearer than it had ever been, but complicated negotiations had yet to be concluded between the Federal and State authorities, and the American Congress had yet to give the project the final seal of its approval. Into a situation that

241

still called for infinite tact and discretion, Robert Saunders, working largely behind the scenes and despite the exuberant verve and forthrightness that had marked his career as Controller and Mayor of Toronto, had brought the important qualities of statesmanlike negotiation.

It should go without saying, of course, that these accomplishments were not Saunders' alone. Sharing the credit with him was the supremely loyal and efficient organization that had been built up by his predecessors over almost fifty years—administrators, engineers, technicians, construction workers, operators, linemen, and municipal employees. All, from the general manager to the lonely overseer walking patrol in far-off Patricia, had played their parts.

In recognition of these facts—and to give substance to them—the decision was reached in 1950 to re-name a number of the larger generating and transformer stations in honor of members of the Hydro staffs who, through their association with the Provincial and municipal organizations, had made distinguished contributions to the success of the public-ownership enterprise and to that of engineering in general. Henceforth the pioneer Queenston-Chippawa plant would be known as Sir Adam Beck-Niagara GS No. 1. Other stations rechristened at the time were the E. V. Buchanan TS, Westminster; Ross L. Dobbin TS, Peterborough; Richard L. Hearn GS, Toronto; Otto Holden GS, La Cave, Ottawa River; J. Clark Keith GS, Windsor; A. W. Manby TS and Service Centre, Islington; George W. Rayner GS, Mississagi River; and R. H. Martindale TS, Sudbury.

Speaking at the ceremonial opening of the handsome Des Joachims GS on the Ottawa River in the early summer of 1950, the Premier of Ontario gave further affirmation of the progress of the period. The power shortage, declared Mr. Frost, was already ended. Hydro was then serving 1,132 municipalities comprising 27 cities, 116 towns, 147 villages, 184 police villages, 639 townships, 9 improvement districts, and 10 mining townsites—from 64 hydro-electric and 7 fuel generating stations, and from various sources through purchase.

But Mr. Frost's statement was somewhat premature. In June 1950 the Korean war stepped up demand so that, despite favorable water conditions and the increase of generating capacity, supply in the Southern Ontario System remained short of primary requirements; some industrial customers were forced to operate with restricted loads.

In addition to using all of the energy available under the principal purchase-power agreements, the Commission now took delivery of excess energy from the Canadian Niagara, Beauharnois, Gatineau, Maclaren-Quebec, and Ottawa Valley Power companies, and purchased power from customers who had diesel and steam-electric stand-by units.

At Des Joachims GS, seven out of eight units were now in service, yielding 315,000 kilowatts, and at Chenaux GS, also on the Ottawa River, two out of eight units were producing 30,000 kilowatts. In the Thunder Bay System, both units at Pine Portage GS on the Nipigon River were in operation, producing 60,000 kilowatts and helping to meet all primary demands including those of the Rainy River district. In the Northeastern Region all districts had been completely integrated and a tie-line gave connection with the new plants on the upper Ottawa River. At the George W. Rayner GS on Mississagi River, both units were in service, adding 47,000 kilowatts, and power was being purchased from Quebec Hydro.

The rural five-year plan of 1946 had been completed and had passed its original objective, raising the average aggregate peak load by 151 per cent. Transmission lines had been increased by 80 per cent and customers by 135 per cent above the targets set at the beginning of the five-year plan. Average cost to farm service customers had been reduced from 2.11 cents per kwh in 1944 to 1.85 cents per kwh in 1950.

To meet demands, the Commission decided to install three additional generating units at its two principal thermal stations: two at the Richard L. Hearn GS at Toronto and one at the J. Clark Keith GS at Windsor. The installations were made necessary by the continuing failure of authorities in the United States to coöperate with Canada in a succession of proposals for the development of the St. Lawrence and the division of boundary waters, which required action on the part of both Congress and the State of New York, with the concurrence and approval of the International Joint Commission.

It will be recalled that in the early 1920's, the International Joint Commission had recommended the appointment of the Joint Board of Engineers, which in 1926 reported favorably on the Canadian-U.S. development of the St. Lawrence River for navigation and hydroelectric purposes. Largely on the basis of the Board's recommendations, the two countries in 1932 signed the St. Lawrence Deep Waterway Treaty to provide for their development of the resources of the Great Lakes Basin in the interests of both deep water navigation and power. Despite the approval of President Herbert Hoover and his successor, Franklin D. Roosevelt, formidable opposition toward the Treaty was organized in the United States, and it was rejected by the Senate in 1934.

Further studies followed, and impelled by the power needs of World War II, Canada and the United States, with the same objects in view, negotiated still another treaty—the Great Lakes-St. Lawrence Basin Agreement—in 1941. This instrument, like its forerunners, was submitted to the Senate for ratification but failed to secure the necessary two-thirds majority, and despite President Truman's pressing endorsement, remained unratified.

Fortunately, at this point, a different attempt to reach international agreement to develop another valuable shared resource was fully successful. Largely on the initiative of Ontario Hydro and electric utilities in the State of New York, working through their respective governments, a new Niagara Diversion Treaty was negotiated. Whereas former agreements had specified the maximum flows that could be diverted for power development, the new agreement specified the minimum flows that must remain undiverted and hence available for scenic display. The minimum flow was to be at its greatest during daylight hours of the peak tourist season. Apart from the tourist season, under the new treaty, the water remaining after specific provision for scenic purposes would be divided between Canada and the United States, to provide a maximum use of the available flow for the generation of power. The treaty also provided for the Niagara Remedial Works.

The treaty was signed by President Truman in March 1950 and was ratified by the Senate in October of the same year. It then became possible for Ontario Hydro to proceed with a huge new development at Niagara which would install 1,370,000 kw of generating capacity—46 per cent more than would be the Ontario share in the blocked St. Lawrence Power Project and more than three times as large as the Sir Adam Beck-Niagara GS No. 1, for many years the world's largest.

The Sir Adam Beck-Niagara GS No. 2 was to be built alongside the existing plant at Queenston. The water for its sixteen units would likewise be diverted from the Niagara River near Chippawa; however, for most of the distance to Queenston, it would be conveyed through a pair of pressure tunnels, 45 feet in finished diameter, carried far below the city of Niagara Falls, rather than through an open-cut canal. The excavation of these tunnels, each more than five miles long, and of the two and a quarter mile canal from their exits to the forebay, would provide some of the rock for the banks of a pumped-storage reservoir with an area of 750 acres. The water pumped into this reservoir during hours of low demand for power would be released to generate electricity when demand soared to its peak. The same six units would serve first as electric-powered pumps and then as generators. This novel arrangement for developing the maximum of electrical energy from all the available water would several years later be a feature of the great Lewiston development on the U.S. side of Niagara. Construction of the Sir Adam Beck-Niagara GS, No. 2 actually began in December 1950, less than three months after the U.S. Senate had ratified the Niagara Diversion Treaty.

Despite this mark of progress on Canada's part, domestic controversy in the United States respecting the rights and powers of the Federal Power Commission on the one hand and the State of New York on the other, had further delayed progress with the St. Lawrence development envisaged in the unratified agreement of 1941. At this juncture, Prime Minister Louis St. Laurent came to one of the most dramatic decisions in his country's history. Grown restive with delay and frustration, and impatient with the obstructionist tactics of American business interests and the U.S. Congress, Mr. St. Laurent, although himself a native of Quebec where the Seaway had failed to arouse marked enthusiasm, told his Minister of Transport, Lionel Chevrier, in the summer of 1950, that, "We should build the seaway alone. I think the Americans should be made aware of our determination to get the seaway built."

The quotation is from Mr. Chevrier's graphic work, *The St. Lawrence Seaway*, in which he relates his experiences as Minister of Transport from 1945 to 1954 and as President of the St. Lawrence Seaway Authority from 1954 to 1957, when he resigned in order to re-enter politics. Born in Cornwall, Ontario, he played as a boy in the old St. Lawrence Canal, and as Minister of the Crown made the first announcement that Canada could, and if necessary would, make the completion of the seaway its undivided responsibility.

The American reaction was one of astonishment, then incredulity. Was this a determined government policy or simply Mr. Chevrier's own personal opinion? The answer was given shortly when the Government at Ottawa let it be known that Canada was prepared to proceed with an all-Canadian waterway as far west as Lake Erie, once the means had been found to have the power works constructed concurrently in the international section of the St. Lawrence River. By December 1951 the Canadian Parliament had passed the St. Lawrence Seaway Authority Act and the International Power Development Act, the first authorizing the construction of navigation works on the Canadian side of the river from Montreal to Lake Ontario, and the second authorizing Ontario Hydro to join with a designated U.S. power authority in constructing the necessary hydro-electric power works in the International Rapids section of the river. The two bills received unanimous consent, the three leaders of the Opposition—George Drew of the Progressive

Conservative Party, M. J. Coldwell of the Coöperative Commonwealth Federation, and J. H. Blackmore of the Social Credit Party—all giving them their wholehearted approval.

Very largely as a result of Canada's sovereign intransigence, a profound change in American opinion, coupled with the development of Labrador's iron resources by American capital, broke the log jam that had persisted for more than a generation and, in 1952, the Canadian and American governments submitted applications to the International Joint Commission for approval of a combined power development, including dams and necessary auxiliary works. In line with these developments, the U.S. Federal Power Commission in 1953 granted a fifty-year license to the Power Authority of the State of New York for the development of the American half of the power project. Although the order granting this license was contested in the U.S. courts, and was not validated until 1954, the way was now cleared for the coöperative planning of the power enterprise by Ontario Hydro and the New York Power Authority.

With the forty-year old dream on the threshold of realization, the appointment of Robert Moses as Chairman of the New York authority provided Robert Saunders with a colleague whose record in the municipal affairs of New York City was not unlike that which Saunders himself had made as Mayor of Toronto. Alike in personality, energy, and the ability to get things done, the two men constituted a brilliantly efficient team.

In the meantime, while the Commission and its Chairman had been mainly concerned with the major policy decisions involved in negotiations on several fronts, the day-by-day operations of the organization had gone forward under the direction of R. L. Hearn as General Manager and Chief Engineer, Otto Holden as Assistant General Manager —Engineering, and A. W. Manby as Assistant General Manager—Administration.

In 1951 the Ontario Hydro-Electric Advisory Council which had been established in 1944 but never constituted, was increased from five to nine members, to provide a cross-section of public opinion on Hydro affairs, and more particularly with respect to the somewhat delicate customer relations precipitated by the 25 to 60-cycle conversion program. In addition to representatives of the OMEA, AMEU, commerce and manufacturing, were included those of labor, journalism and agriculture. For the first time Ontario Hydro's principal domestic users received recognition in the appointment of Mrs. Marjory Hamilton, Mayor of Barrie and a well-known leader in women's affairs. That same year the Rural Telephone System Act was also passed to improve, extend and coördinate rural telephone systems by means of Hydro's assistance.

Despite the step-up of Hydro's construction program, expanding demand continued to exceed available generating capacity. Restrictions on the use of primary power were avoided only by limiting deliveries of "at will" and interruptible power. Since the inception of Hydro's tremendous expansion program in 1945, the dependable peak capacity of all systems had been increased by more than one million kilowatts at a cost of $651,054,956, with an additional $322,155,536 planned and approved.

Due to the impasse on the St. Lawrence, however, Hydro's progress swirled and eddied, seeking other sources of desperately needed power. From 1950 through 1955 the familiar pattern of expansion continued on all fronts. The number of rural and urban customers increased, as did their individual and total electrical needs, to serve which additional generating and transformer units were installed and new power sites surveyed. Not only were power sources consolidated within systems but the systems themselves were integrated with each other by interconnections. Considerable progress was made in frequency standardization changeovers and the Legislature coöperated by providing authority for the successive steps required to achieve a single vast power pool.

By the end of 1952 the engineering teams engaged in the gigantic standardization changeover had reduced the geographic area of the "25-cycle island" from 12,000 square miles to 7,000. The growth of population and industry in the province, combined with the increasing use of frequency-sensitive equipment, added substantially to the magnitude and complexity of the operation, whose costs, calculated in 1947, were soon outdated by rising prices and extensions to the basic plan.

In 1952 the Commission's dependable peak capacity reached 3,353,350 kilowatts—73.1 per cent greater than in 1945. Additional units were brought into operation at the Richard L. Hearn GS, the J. Clark Keith GS, and the Otto Holden GS. Construction at the powerhouse site of Sir Adam Beck-Niagara GS No. 2 was well advanced. The Thunder Bay System was now consolidated with the Northern Ontario Properties, divided into the Northeastern and Northwestern Divisions for operational purposes.

By 1953 the distribution of primary energy was in the proportion of 55 per cent to the municipalities, 37 per cent to the large industrial customers, and 8 per cent to rural power districts. Each domestic customer was using an average of 4,404 kwh, compared with 2,454 kwh in 1945. Significantly, the average kwh cost was 1.155 cents as compared with 1.074 cents in 1945 and 1.259 cents in 1939.

What the Commission in 1953 described as its "all-out effort . . . to keep abreast of the mounting power demands" reflected the growing level of prosperity in the province, large-scale expansion of industry, and increased farm output. Since the initiation of the expansion program in 1945, capacity had increased by 84 per cent. The demand for primary power was up in all systems, but the relationship between demand and production in 1953 was the most satisfactory that had been reached since 1945. This was the more remarkable in that during the year there was an unusually severe shortage of water in the Ottawa River watershed, necessitating expanded output by the large fuel-electric stations, where the major increases in generating capacity were made during this period.

Also in 1953 the third and fourth units at Pine Portage GS were under construction and at Manitou Falls on the English River, initial construction began on the largest, up to that time, hydro plant designed for remote radio-controlled operation—a three-unit generating station to produce 42,100 kilowatts by 1956. In the Northeastern Division production in excess of primary power requirements was supplied as secondary energy to the paper industry and surplus energy could also be advantageously disposed of in the Southern Ontario System where interconnection had been made with the Detroit Edison Company at Windsor to assist with the increased 60-cycle load.

The interconnections between the Southern Ontario System, the Detroit Edison Company and the Northern Ontario Properties proved of great service in 1954, when the Richard L. Hearn GS in Toronto was forced to close down for some time following an unusual mechanical failure and a subsequent fire. In the emergency the Commission bought power from the Detroit company and from the Niagara Mohawk Power Corporation at Niagara Falls, New York.

In the rural districts 1954 was marked by consolidation and improvement of existing facilities rather than expansion. With 85 per cent of Ontario's farms using electricity, the need now was for improved transmission facilities to meet the increased loads inevitably following initial delivery of power to farmhouses that hitherto had known only the kerosene lamp and wood stove. As in the municipal areas, the erosion of the 25-cycle island continued as rapidly as conversion crews could move from one location to the next.

At the close of still another year of unprecedented effort, Hydro was able to report, for the first time since the depression of the 1930's, a reasonable power reserve in all systems. During the twelve-month period there had been achieved a 16 per cent increase in power resources, a 7 per cent increase in the amount of energy generated and purchased, and a 5.6 per cent increase in the number of customers served or scheduled to receive power in the near future.

The outstanding events of 1954, however, were two memorable ceremonies, the one to mark the official opening of the Sir Adam Beck-Niagara GS No. 2 by Her Royal Highness the Duchess of Kent; the other, the long deferred inauguration of the St. Lawrence Power Development and Seaway Project at Cornwall, Ontario, and Massena, New York. Appropriately, since the need for power had provided the original incentive, it fell to the two power authorities of Ontario and New York in the persons of Robert Saunders and Robert Moses to officiate as the representatives of their respective countries.

Even before the opening ceremonies were concluded at a banquet held in the Cornwall Armories, construction crews and earth-moving machinery were already working on engineering plans that had been approved and held in readiness for several years. Before many months passed, however, a tragic accident was to deprive both Hydro and the St. Lawrence power development of the drive and gusto of its most dynamic champion, Robert H. Saunders.

Late Friday night, January 14, 1955, after addressing a meeting at Harrow, Saunders flew out of Windsor in a twin-engined Mallard. After midnight the plane came in for a landing at Crumlin Airport, London. Its wings and fuselage had become loaded with ice, forcing it to crash-land in a field 4,800 feet short of the runway. Mr. Saunders was seriously injured and was taken to Victoria Hospital, London, where, after a twenty-nine hour fight for life, he died at 5.25 a.m. on January 16 at the comparatively early age of fifty-one.

27

On January 24, 1955, Richard L. Hearn was appointed to succeed Robert H. Saunders as Chairman of the Commission, at that time enlarged by an amendment to the Power Commission Act, which increased membership to a total of six Commissioners. Dr. Hearn graduated from the University of Toronto in 1913 with a degree in Applied Science and began his career with Ontario Hydro the same year as a designer of the Wasdell Falls power plant. Appointed Assistant Engineer on Construction in 1918, he supervised design and drafting on the Queenston-Chippawa project and on several smaller projects including an extension to the Ontario Power GS. From 1921 to 1942 Dr. Hearn occupied executive engineering posts with various private companies in the United States and Canada, returning to Ontario Hydro in the latter year as Executive Assistant to the Chairman. He was subsequently loaned as Chief Engineer in charge of construction of the Polymer Corporation, a synthetic rubber plant at Sarnia, and later as Canadian Technical Advisor to the Public Utilities Division of the Combined Production and Resources Board at Washington, D.C. In 1945 he was appointed Ontario Hydro's Chief Engineer— Design and Construction and, in 1947, General Manager and Chief Engineer. Among the many honors conferred on Dr. Hearn in recognition of his contributions to Hydro and Canadian engineering in general are the degree of Doctor of Engineering, bestowed by the University of Toronto in 1952, and the coveted Julian C. Smith Medal, awarded by the Engineering Institute of Canada in 1954. In addition, the Association of Professional Engineers of Ontario presented Dr. Hearn in 1956 with the Professional Engineers' Medal and citation "for his distinguished accomplishments in the sphere of hydro-electric engineering and for his contribution to mankind in furthering the application of nuclear fission in peacetime uses."

On January 24, the same day that Dr. Hearn was named Chairman of the new Commission, A. W. Manby was appointed General Manager, and Dr. Otto Holden became Chief Engineer. George H. Challies and W. Ross Strike continued as Vice-Chairmen and E. B. Easson as Secretary. In May 1955 Mr. Challies retired, after nearly twelve years on the Commission, to become Chairman of the newly established Ontario St. Lawrence Development Commission, a body with the responsibility of preserving and enhancing the scenic attractions and historic associations of the area adjacent to the St. Lawrence Power development. He was succeeded at Ontario Hydro by the Hon. William Hamilton. An additional Commissioner was also appointed at this time, Lieut. Colonel A. A. Kennedy, DSO, President of OMEA, and distinguished both as a soldier and a businessman. In his native Owen Sound he had long been active and influential in civic and Hydro affairs. Later in the year, William Hamilton, who had represented Wellington South, lost his seat in the general election. In August 1955 he was replaced on the Commission by the Hon. W. K. Warrender, formerly Minister of Planning and Development.

A. W. Manby joined the staff of Ontario Hydro in 1921. Born at Niagara Falls, Ontario, he received his early education there, and graduated from the University of Michigan with the degree of B.SC. in 1921. He interrupted his university career from 1917 to 1919 to serve as Flying Instructor with the Royal Flying Corps, and later as Flying Officer with the 46th Squadron of the Royal Air Force in France.

Upon graduation Mr. Manby entered the service of the Commission and spent a year with the Construction Department during the building of the Sir Adam Beck-Niagara GS No. 1 at Queenston. From 1922 to 1931, he was the chief operator of this key station, following which he was appointed Superintendent of the Chats Falls GS. Three years later he became Chief Operator, Niagara System, stationed at Toronto, serving in this capacity until his appointment in 1938 as Assistant to the Chief Operating Engineer at Toronto. In 1941 Mr. Manby became Assistant to the Chief Engineer, a post he held until March 1947, when he was named Assistant General Manager—Administration. In 1950, in recognition of this record of distinguished service, Hydro's great service centre and transformer station at Islington were named after him.

Dr. Holden graduated from the University of Toronto and joined Ontario Hydro in 1913 as a draughtsman. In 1924 he became the deputy head of the Hydraulic Department. His quality became apparent through work on the Queenston-Chippawa development and on the Nipigon River project at Cameron Falls. He superintended the design and construction of the first complete development of an interprovincial power site on the Ottawa River—the Chats Falls GS, which was started in 1929. In 1937 and for ten years thereafter he was Chief Hydraulic Engineer; then Assistant General Manager—Engineering, in which capacity he administered Ontario Hydro's Planning, Engineering, Construction, Research, and Supply Divisions. He carried a great deal of responsibility for the Commission's tremendous post-war expansion program, including the building of the three major generating stations on the Ottawa River: Des Joachims, Chenaux and La Cave. The latter station was re-named in his honor and became the Otto Holden GS.

Dr. Holden was a member of the International Committee in charge of the design and construction of the remedial works in the Niagara River which were designed both to preserve and enhance the beauty of the Falls and to make better use of the water in the production of power. Representatives of the Snowy Mountains Hydro-Electric Authority of Australia visited Ontario to inspect Hydro installations and were so impressed that they invited Dr. Holden to examine and report on the Authority's plans for a hydro-electric project in southeastern Australia.

One of the Commission's objectives had been to create a power reserve of 12 per cent. Dr. Hearn's exertions to that end increased peak capacity by 9.6 per cent in 1955, but meanwhile demand rose by 14.3 per cent and the situation was aggravated by unfavorable water conditions. The Commission was now operating a total of sixty-five hydro-electric and five thermal-electric stations, but more would be needed, and there remained few undeveloped hydro-electric sources. Would nuclear power be available in time to meet future demands?

For an answer to this question the Commission was relying upon the experience that would be gained by the operation of an experimental generating station at Chalk River, a pilot station scheduled for construction by a partnership formed of Ontario Hydro, Atomic Energy of Canada Limited and the Canadian General Electric Company Limited. The Commission's experts were participating in this experiment so that if during the next decade nuclear power stations became economical for base load operations, Ontario Hydro would be prepared with the engineers and the technical knowledge needed for efficient exploitation of this new and revolutionary source of power.

Meanwhile, the still-available water-power resources were being rapidly developed. Five more units were in service at Sir Adam Beck-Niagara GS No. 2, and the thirteenth and fourteenth units were under construction, with two more planned for construction in 1956, these additions to be worked in conjunction with the pumped-storage generating station. At the Richard L. Hearn GS in Toronto an additional 200,000-kilowatt unit was planned, and on March 15 and October 13, 1955, the two units that had been damaged by fire were returned to service. In the Northern Ontario Properties, interchange facilities between the Southern Ontario System and the Northeastern Division were improved, while in the Northwestern Division at Pine Portage GS the fourth unit was in operation, having been placed in service at the end of 1954. At Manitou Falls GS on the English River, four units were already in service and a fifth was planned; on the same river the development of a site at Caribou Falls was under discussion. Other new power resources planned or under construction included extensions to the Cameron Falls GS and Alexander GS on the Nipigon River and a 54,000-kilowatt generating station at Whitedog Falls on the Winnipeg River near Kenora. In that area there was a rapid increase in load which was expected to continue; in consequence Ontario Hydro was considering the possibility of interconnection with the Manitoba Hydro-Electric Board.

In 1956 Hydro celebrated its Golden Jubilee—the fiftieth anniversary of the passage of the Power Commission Act by the Ontario Legislature. The Act had received Royal Assent on May 14, 1906, and its first Chairman and Commissioners were appointed on June 7. During the succeeding half-century Hydro had become an example to the world of democracy at work. The principal official celebrations were held in Kitchener, formerly Berlin, where the first public-ownership meetings had been held, and which had been the first of the fourteen original coöperating municipalities to receive Hydro power. At nearby St. Jacobs a slender 20-foot memorial obelisk of Queenston limestone, surmounted by a perpetually lighted lamp, was unveiled to the memory of Hydro's founding father, Elias Weber Bingeman Snider, ("E.W.B."), who had presided at the crucial organization meetings of June 9, 1902 and February 17, 1903. The unveiling was performed by a member of another notable Waterloo family, the Hon. Louis O. Breithaupt, Lieutenant-Governor of Ontario. Presiding at the ceremony was the President of OMEA, Gordon H. Fuller, who made a speech of tribute to "E.W.B." The gathering of about four hundred people included members of the Snider family, representatives of Ontario Hydro and the municipal electrical utilities, and municipal officials.

In the evening, at the Kitchener Memorial Auditorium, eight hundred guests attended a gala Jubilee dinner at which J. W. Washburn, Chairman of Kitchener Hydro, presided. Among the speakers were Lieutenant-Governor Breithaupt, Dr. Marcus Long of the University of Toronto, and Ontario Hydro's two Vice-Chairmen, W. K. Warrender and W. Ross Strike. Ontario Hydro Commissioner A. A. Kennedy presented framed two-dollar bills to representatives of each of the original participating municipalities in remembrance of the two-dollar subscription which they had paid for the historic banquet at the Walper House in Berlin on June 9, 1902. A pageant with music dramatized Hydro's fifty-year history and to mark the anniversary, the Commission issued a special Golden Jubilee number of its publication, *Ontario Hydro News.*

During the year there were further changes in the membership of the Commission. The first of these was the retirement on October 31 of the Chairman, Dr. Richard L. Hearn, who, however, maintained his Hydro connection as consultant. Dr. Hearn was succeeded as Chairman by James S. Duncan, CMG, LL.D.

A leading Canadian industrialist with a wide and varied experience in many parts of the world, Mr. Duncan was born in Paris, France, where his Scottish father was Manager for France of Massey-Harris, the Canadian manufacturer of farm machinery. Educated in France, at sixteen he joined Massey-Harris in Berlin, Germany. From office-boy there to factory-hand in Canada and then to the sales staff, his rapid rise in Massey-Harris was interrupted in 1914 when he enlisted in the Royal Field Artillery. After gaining promotion from the ranks to a captaincy and a Mention in Despatches, he rejoined the European staff of Massey-Harris in 1919.

He progressed steadily to the post of European General Manager in 1931, General Manager of the Argentine Division in 1933, and General Manager at Head Office in Toronto in 1935. In 1941 he became President of the company, and in 1949, Chairman of the Board and President.

Prior to his retirement in June of 1956, his expert guidance of the enterprise to which most of his life and great energy had been devoted had earned him international recognition, illustrated when the U.S. National Sales Executives named him "Canadian Businessman of the Year," for 1956, and when Dartmouth College honored this "pioneer among the growing group of far-sighted business-statesmen who bring both vision and daily reality to the free world's cause," with a degree of Doctor of Laws in 1957.

Early in World War II, Mr. Duncan had been called to Ottawa as Acting Deputy Minister of National Defence for Air. His assignment was to organize the British Commonwealth Air Training Plan. His success in this challenging task, combined with many other notable wartime activities earned him high honors from the governments of Great Britain, France and Norway.

After the war, Mr. Duncan interested himself actively in promoting Canada's export trade. In 1949 he played the leading part in the formation of the new Dollar Sterling Trade Council and became its first Chairman. In November 1957 he was named Deputy Chairman of the Canadian Trade Delegation to Great Britain. A long-established interest in education culminated in his appointment in 1956 as Chairman of the National Conference on Engineering, Scientific, and Technical Manpower.

The appointment of James S. Duncan to the chairmanship of Ontario Hydro on November 1, 1956, may be said to have added a new dimension to the Commission. The industrialist with the sales background, the expert in business organization, the multi-lingual cosmopolitan, the dynamic and eloquent public speaker, the leader in community and national causes, would respond to the challenges of a new era in the history of Hydro, challenges arising from the transition to fuel-electric generation, the competition of natural gas, and the rushing progress of automation and electronics in industry.

The Hon. W. K. Warrender resigned on November 1 to become Provincial Minister of Municipal Affairs and was succeeded as government representative by the Hon. T. Ray Connell. On the latter's appointment as Minister of Reform Institutions in 1958, Robert William Macaulay was made Second Vice-Chairman. Mr. Macaulay was called to the Ontario Bar in 1948 and six years later, at the unusually early age of thirty-three, was created a Queen's Counsel. Co-author of a legal textbook, and the author of several published articles on legal subjects, he has been associated with the well-known Toronto legal firm of McLaughlin, Macaulay, May and Soward, and lectures at the University of Toronto Law School.

First elected to the Provincial Legislature in 1951 to represent the riding of Riverdale-Toronto, Mr. Macaulay was re-elected in 1955 and again in 1959, with a substantial majority. He has taken an active role in the work of several select committees appointed by the Ontario Government to study specific subjects, having served as Chairman of one committee and as a member of four others. In May 1959 he was named Ontario's first Minister of Energy Resources, though still retaining his position in Ontario Hydro.

D. P. Cliff of Dundas, President and Secretary-Treasurer of OMEA, was appointed a Commissioner on November 1, 1956. Mr. Cliff, for many years a Commissioner of the Dundas Public Utilities Commission and its Chairman for four years, had served on many public bodies and on the Town Council of Dundas, his native place, of which he had been successively Deputy Reeve, Reeve and Mayor.

During 1957 Canada felt the first tremors of a general economic recession and Hydro, with the demand curve tending to level off, experienced no power shortage. A well-integrated thermal program was under way and nuclear-electric power was on the horizon. To ensure the continued load growth on which the maintenance of low rates depended, the Commission, in coöperation with the municipal utilities, embarked on a sales promotion campaign with particular emphasis on home and farm use, while the municipal utilities took measures to improve service and customer relationships.

More large fuel-electric generating stations were planned for Hydro's next ten-year development: Lakeview GS on the Toronto Lakeshore; others east and west of Toronto; and Thunder Bay GS at Fort William. At Niagara, completion of the Sir Adam Beck-Niagara GS No. 2 could be expected in 1958. In the Northern Ontario Properties work went forward on seven new hydro-electric installations which would increase capacity by 254,000 kilowatts.

During most of 1958 the increase of Ontario Hydro's power load fell far below the long-term average annual rise of 6.5 per cent. By December, however, recovery was well under way. Rates remained unchanged despite the inflationary trend of labor and material costs, and rising taxes. Ontario Hydro's resources were increased in 1958 by 19 per cent—the largest additional capacity ever recorded in a single year. This enabled the Commission at last to provide the 12 per cent power reserve which had been an objective ever since the end of the war.

In July 1958 the Robert H. Saunders-St. Lawrence GS came into service and in August the Sir Adam Beck-Niagara GS No. 2 was completed. This, with the exception of some hydro-electric sites still to be harnessed in Northern Ontario, exhausted Ontario's water-power resources within close reach of large load-centres. In the future the Commission would have to rely increasingly on thermal generation, entailing the importation of large tonnages of coal, until such time as the development of nuclear power would make it possible to utilize Canada's abundant uranium resources. During the year the AECL's Nuclear Power Plant Division was engaged in the design of a 200,000-kilowatt nuclear power plant, to be in operation by 1965 at latest. Meanwhile, at the 20,000-kilowatt Nuclear Demonstration Plant near Chalk River, important new engineering developments were incorporated and the plant was scheduled for completion in 1961.

Calculations of Hydro's future power load were affected in 1958 by the appearance of a new competitor, natural gas. Although the increasing use of the new fuel would not significantly affect peak demands, it could cut materially into off-peak demands, in which case revenues of the municipal systems would be seriously lessened without compensating reductions in their capital or operating costs. This apprehension motivated the energetic promotional campaign begun at this time, emphasizing the particular advantages of electric water-heating and house-heating and reiterating, through almost every possible medium of advertising, the economical, labor-saving and comfort-giving aspects of electrical service in the home. In addition, a mobile display of electrical appliances was set up in a 42-foot trailer to visit fairs, exhibitions and other public gatherings. The costs of this campaign were to prove amply justified by subsequent increases in off-peak domestic consumer demand.

That Hydro sells its product at cost may give it a certain advantage from the point of view of public relations, and these relations are indubitably strengthened by its partnership with the municipal electrical utilities and by the two organizations which represent and implement that partnership—the Ontario Municipal Electrical Association and the Association of Municipal Electric Utilities.

The OMEA—composed of municipal utility commissioners elected by the people (or appointed by an elected body)—represents a true cross-section of the public and the combined voice of more than 70 per cent of all Ontario Hydro customers in matters of policy. The AMEU, representing the technical and managerial staff of the local utilities, acts as an advisory group to the OMEA and to Hydro on all technical and operational problems.

In applied research, Ontario Hydro's achievements over fifty years rival those of any other public utility. In certain fields, particularly in the utilization of concrete, Ontario Hydro engineers have acquired international pre-eminence. The scientific mixing of concrete, replacing the former method of using arbitrary proportions, was pioneered by Ontario Hydro's research laboratory. Among Ontario Hydro's contributions to the technique of concrete construction are the equipment developed to speed up the testing of concrete's reaction to cycles of freezing and thawing, and the soniscope, an instrument employing ultrasonic principles to test the condition of concrete structures. Among the many techniques and special instruments invented by the laboratory for use in the construction and testing of generators, transmission lines, transformers and other electrical installations are the portable bolometer, which measures infra-red radiation from transmission-line joints and indicates any deterioration due to excessive operating temperature, and the linascope, an instrument utilizing radar principles to locate short circuits in faulted power lines. Recently, the laboratory's

258 : THE PEOPLE'S POWER

studies of sprays used in brush control along rights-of-way led to the development of methods of testing the concentration of toxic chemicals in spraying solutions which have proved of substantial value to the insecticide industry and to agriculture. Current research and pioneering effort are directed toward the application of electronic computers to determine which generating station or stations should be called upon to supply most economically and effectively those sudden demands for extra power that inevitably, if unpredictably, occur at frequent intervals.

THE SEAWAY AT LAST

28

On May 13, 1954, in Washington, President Eisenhower signed the Wiley-Dondero Act authorizing the participation of the United States in the construction of the St. Lawrence Seaway. The great international power project for which a succession of Ontario Hydro chairmen, from Adam Beck to Robert Saunders, had argued and fought for over thirty years, officially started on August 10, 1954. Now at last the St. Lawrence Seaway and Power Project was to emerge from the dreams of its prophets and the blueprints of its engineers as one of man's most consequential geographic transformations.

"Across the river we have held hands," said Senator Wiley. "Now we cannot part. We are one in a great adventure—to build for the future of America."

The hands of both nations, represented by the engineering directors of the project, had long been poised and waiting for the scratch of President Eisenhower's pen. The plans of the power project had in fact been agreed upon six years before.

Now at long last the U.S. Government had decided. Everything was ready. On both sides of the international boundary, brigades of giant bulldozers marked time, panting and snuffling, while the Canadian and American authorities met to assign and program their respective portions of the great task. Lionel Chevrier, President of the St. Lawrence Seaway Authority, in his book, *The St. Lawrence Seaway*, has revealed some of the difficulties and embarrassments encountered at these early conferences, when the American officials appeared to be under the impression that they alone held full responsibility for the entire project. They would come to the meetings with a unilaterally compiled agenda, much to the annoyance of some of their Canadian colleagues, who found it necessary from time to time to remind them that the Americans were not in charge of the Canadian section of the undertaking.

259

Perhaps helped by these reminders, the Americans in the end coöperated with exemplary good will and effectiveness. Both the Canadian and the American sections of the power project went forward uninterruptedly, without exceeding the advance estimates of cost, and with no major divergences from the plans which engineers on both sides of the border had been drawing and re-drawing during thirty years of frustrated discussion. During the last decade of this period Ontario Hydro and the Power Authority of the State of New York were in constant consultation, planning precisely the structures needed in the development and the materials, quantities, machines and work schedules. For the power project they planned three principal structures: two adjoining powerhouses, two massive dams, and a system of dykes on both sides of the river to pond the once-turbulent currents of the St. Lawrence. Since these plans required only minor modifications during the course of construction, they constitute in effect a description of the completed project as it stands today.

Adjoining powerhouses, built at the international boundary, span the channel between the eastern end of Barnhart Island and the Canadian shore at Cornwall, Ontario. They are 162 feet high and 3,300 feet long overall. The structures in themselves act as a gravity dam. Power is produced by thirty-two generating units, sixteen in each plant, with an installed capacity of 1,880,000 kilowatts. The powerhouses are of a modified outdoor type, lacking a conventional superstructure over the generators, but with removable covers to protect the units. The architecture is simple and functional. Facilities are provided for visitors to view the structures. The powerhouse on the Canadian side is named after Robert H. Saunders, who led the final drive to get construction started. The powerhouse on the American side is named after Robert Moses, as Chairman of the Power Authority of the State of New York, in honor of his public services.

Upstream from the generating station, curving gracefully for 2,960 feet between the New York mainland and a point near the head of Barnhart Island, the Long Sault Dam helps to exploit the drop in water level along the 125-mile stretch between the east end of Lake Ontario and the powerhouse, and by-passes water not needed for power production. In combination with the downstream power dam, the Long Sault Dam maintains the head of water necessary to turn the turbines, with a discharge capacity far in excess of the maximum flow of the river. The spillway is equipped with thirty 50-foot-wide lift gates, and winter heating keeps the gates and guides free of ice.

The Iroquois Dam, twenty miles upstream from the Long Sault Dam, is an impressive series of piers and sluiceways 1,980 feet long and 67 feet high. It spans the main river channel between Iroquois Point on the Canadian side and Point Rockway on the United States mainland. It controls the outflow from Lake Ontario in accordance with the provisions of the International St. Lawrence River Board of Control.

On the Canadian shore north and west of the powerhouses a dyke three and a half miles long, flat-topped and with sloping sides, contains the headpond created by the Long Sault Dam and the powerhouses. On the American side is similar dyking. To build the Canadian dykes required about five million cubic yards of earth and about 300,000 cubic yards of stone along the sloping sides. The entire project cost the two power agencies over $600,000,000.

Deep under the water of the lake lie sites of the farms and villages, churches, cemeteries, roads and railways of a countryside forever submerged. Seven communities and part of an eighth were affected by the flooding: the villages of Iroquois, Aultsville, Farrans Point, Dickinson's Landing, Wales, Moulinette, Mille Roches, and one-third of the town of Morrisburg. Also involved were 225 farms, 18 cemeteries, 35 miles of highway, and 40 miles of doubletrack railway. Over 6,500 people were due to lose their houses and land when the waters rose. New communities at Ingleside and Long Sault were formed, while whole communities were re-located—Iroquois and part of Morrisburg —in a 20,000 acre area along the north shore of the St. Lawrence. Some buildings were moved bodily from the old place to the new— lifted and carried by great house-moving machines capable of raising 200 tons. Christ Church (Anglican) at Moulinette, a 127-year-old stone structure, was transported intact to its new site by a giant moving machine. Memorials and some graves were also moved. In all, over 2,190 properties were purchased.

The new buildings erected in the new and re-located communities and surrounding townships included 9 schools, 14 churches, 4 shopping centres, 96 multiple-dwelling units, 5 municipal and public buildings and 349 houses privately built. Electricity, waterworks, sewage treatment plants, sidewalks, paved streets and all other necessary amenities were provided. The 35-mile section of Ontario Highway No. 2 was replaced and the 40-mile stretch of doubletrack railway on the CNR mainline between Cardinal and Cornwall was rebuilt in a new location. On the American side, similar transformations were accomplished.

The volume of water required to fill the 100-square-mile headpond, extending from the powerhouses to the Iroquois Dam, was approximately 23 billion cubic feet at an initial elevation of 236 feet above sea level. Soon after the completion of the dams the level was gradually raised to an elevation of 238 feet. By international agreement, this level is to be maintained for up to ten years.

For Ontario Hydro, the most difficult and trying part of the whole vast project was the removal and resettlement of the inhabitants in the basin of the projected reservoir. Some towns and villages along the river had been occupied for over a hundred years by United Empire Loyalist refugees from the American Revolution—Scottish Catholics and French Canadians—and for 3,500 years before that by Indian and earlier races. Before the flood began at Cornwall a team of archeologists dug carefully on Sheek Island to remove relics of the ancient predecessors of the Iroquois. Some old residents refused resettlement and insisted on remaining in their houses until the last moment.

Part of the project was to develop the area adjacent to the shore line and the newly created seaway islands—which were to become public property as a result of Ontario Hydro's activities—into one of Canada's scenic and historical showplaces, with scenic drives, parks, playgrounds, museums, and memorials to one of the great periods in the country's history. Responsibility for this undertaking was vested in the Ontario St. Lawrence Development Commission, an organization set up and financed by the Provincial Government. As Chairman of this Commission the Government appointed the Hon. George H. Challies, who had long been a proponent of the St. Lawrence development.

Preparatory work by Ontario Hydro engineers included the construction of three true-scale models of the river from Prescott down to Cornwall. The first of these, representing the 16.1 miles of the St. Lawrence from Ogdensburg to Leishman's Point, included Galop Island and the old village of Iroquois. It measured 170 feet long and 27 feet across at the widest point. The second, which showed the Ogden Island area and the town of Morrisburg, was 83 feet long and averaged 13 feet wide. The third, representing the dewatering and closure area from Cat Island to a point just below the international powerhouse site, was 146 feet long and a maximum of 40 feet wide, and covered an area of 5,010 square feet. It included miniatures of the two principal structures, the powerhouses and the Long Sault Dam.

These models accurately reproduced to scale a 35-mile stretch of the river, simulating actual river conditions and duplicating the shore line, river-bottom contours, and the currents and eddies of the river flow. Ontario Hydro engineers used them to test and compare different types of construction for the three principal structures—the main dam, the powerhouses, and the Long Sault and Iroquois Dams. They were also used by contractors to determine exact locations, local topography, means of access, and velocities and depths of the river water. In addition to the huge three-section model of the river, smaller models were built to aid in planning details of construction: a transparent plexiglass model of the powerhouse ice chutes which helped greatly to solve the problem of passing the ice without damage to the chute; a model of the Long Sault Dam and its immediate area including piers, sluices, wing-walls, tunnel ports for each of seventeen sluices and removable roll-ways in the remaining thirteen sluices. Models were used to reproduce the Massena intake, the powerhouse water passages, the Iroquois Dam and its adjacent area, and the powerhouse tailrace area. A plexiglass testing flume was constructed and used to develop details of sluice-construction at both dams; also for studies of the energy-dissipating works which were constructed adjacent to the dams. These and other models which were used continuously during the construction of the seaway and power development made possible the saving of millions of dollars and repaid their cost many times over.

On August 10, 1954, at ground-breaking ceremonies conducted by Canada and the United States at Cornwall, Ontario, and Massena, N.Y., the great project moved from the stage of planning to construction. As Chairman Saunders declared, it was "a proud moment in the history of the Canadian people." On the American side, New York's Governor, Thomas E. Dewey, pressed a button which exploded a dynamite charge near the site of Long Sault Dam, where work was to begin at once on the initial cofferdam. A half-hour later on the Canadian side, Prime Minister Louis St. Laurent, Ontario Premier Leslie M. Frost, and Governor Dewey, each dug a shovelful of earth, breaking ground for the construction of the international powerhouse. Afterward, at a luncheon in the Cornwall Armouries, Saunders addressed representatives of the committees affected by the power project. Ontario Hydro, he said, was concerned to plan *with* the people of the area, rather than plan *for* them. As he spoke, trucks were rolling along New York and Ontario roads carrying bulldozers, cranes, drag-line equipment, explosives, cement and steel—all the paraphernalia of the great international construction force that had been assembled to make the St. Lawrence, in the words of the Canadian Premier, "a bond rather than a barrier between the neighbouring nations."

The task of Ontario Hydro and its contractors was greatly expedited by harmonious labor relations, based on an agreement signed early in 1955. It was an extension of the Allied Council principle, successfully applied at the Niagara and other Ontario Hydro construction projects, by which a number of A.F. of L. unions undertook to negotiate jointly through a council of the unions on the job. The agreement covered wage rates and working conditions of all Ontario Hydro contractors' employees.

Ontario Hydro and its contractors concentrated first on construction around the Canadian powerhouse area, and on the two cofferdams to be used in dewatering the powerhouse site. The first dam stretched 500 feet across the north channel of the river from Sheek Island on the Canadian side to Barnhart Island on the United States side; the second, 4,300 feet long, spanned the north channel from the Canadian mainland to Barnhart Island. During 1955 the chief activity was the heavy task of preparing work sites for the construction of the two dams. Most of this work was completed by the middle of 1956, and by the end of that year a good start had been made on the permanent structures. In 1957 construction was far enough along to permit the installation of turbines, generators and other equipment in the powerhouse. Diversion of the Cornwall canal was accomplished in that year along with the construction of the concrete closure structure; also the new forty-mile length of the re-located CNR line between Cardinal and Cornwall. Seventy-five per cent of the channel improvement work was completed as well as about 85 per cent of the St. Lawrence TS, and all but a few miles of the No. 2 Highway between Cornwall and Iroquois. In December the Iroquois Dam was turned over to Ontario Hydro's Regional Operations staff by the United States contractor. The abandoned townsites had been cleared and the complex work of resettlement was almost concluded.

Early in 1958 the last big "pour" at the powerhouse completed all but a final 4 per cent of the concrete work. Ontario Hydro's pre-eminence in concrete construction, the product of three decades of pioneering research, was again demonstrated. Even in severe winter weather, work went ahead with no shutdown. Sand was pre-heated by steam, water was warmed for mixing and chemical heat generated within the mix itself. Wooden forms covered with tarpaulins were used to insulate the setting concrete and steam-fed heaters blew warm moist air over the surface of the curing concrete. One of the American engineers, watching heated concrete being poured into the powerhouse foundation said, "I hope those fellows know what they're doing. Our men won't pour if the temperature goes below zero."

The Canadians knew what they were doing. Project Director Mitchell found it more economical to keep construction going throughout the winter with experienced men rather than to shut down and have to train men again in the spring.

Excellent teamwork characterized the relations of Ontario Hydro's engineering and construction forces with the crews employed by the

New York State Power Authority. Both worked to capacity to maintain the pace of construction which was scheduled to be completed on July 1, 1958—Dominion Day. As the target date neared, the entire area seethed with final preparations. The last boat passed through the locks of the old Cornwall Canal, while Ontario Hydro crews removed lighting and power lines from the area to be flooded. Machinery, lock gates and bridges of the old canal were carted away. Pontoons were placed under bridges to float them to new elevations as the water rose. Technicians made final checks of equipment and divers placed the last of the hermetically sealed steel containers of nitroglycerine into the 25-foot holes that had been drilled into the earth and rock of Cofferdam A.1. The parking places and stands set aside for observers in the dyke area were gay with flags and bunting.

To the west, along twenty-eight miles of shore line, now emptied of the life that had once lived there, other bridges stood ready to span the water that would presently appear below them. Sand had been laid to form tomorrow's beaches. A dock with a new marine supply depot stood in a field awaiting the coming flood.

A control room had been set up in the administration building of Ontario Hydro's powerhouse. At four o'clock, in the dark before the dawn of Dominion Day, a button was pressed. Gates opened in the Iroquois Dam and below them the water began to rise. At seven o'clock police hastily cleared the crowds from the danger area. On the power dam and dyke the waiting spectators heard the count-down relayed to them by loudspeakers.

Out of sight of the spectators, in a log bunker additionally protected by sandbags, Dr. Otto Holden, Ontario Hydro's Chief Engineer, sat crouched beside J. Burch McMurran, Chief Engineer of the New York State Power Authority. At eight o'clock the count-down ended, the men in the bunker pressed a button and tons of earth and rock shot into the sky. Through great gaps in the cofferdam, water poured at the rate of 95,000 cubic feet per second into the headpond above the powerhouse.

From their perch on the top deck of the powerhouse the assembled dignitaries led the applause. Among the distinguished guests welcomed by Ontario Hydro Chairman James S. Duncan and New York Power Authority Chairman Robert Moses were Premier Frost of Ontario, Governor Averell Harriman of New York, General A. G. L. McNaughton, Chairman of the Canadian section of the International Joint Commission, other Commissioners, members of the Federal and Provincial Governments, Hydro officials, executive officers of the OMEA and AMEU, civic officials and members of various organizations concerned with the construction and administration of the seaway and power projects.

At the Dominion Day luncheon given jointly by Ontario Hydro and the Power Authority, Mr. Duncan praised the work of the International Joint Commission, the labor force, and the contractors:

> When the first sod was turned in August, 1954, we stated that the waters would be raised in July, 1958—this has been done; that the first power would flow from some of our units in July, 1958—this will take place; that the project will be completed by early 1960—it will be completed by the end of 1959. But even more remarkable still, away back in 1953 . . . we estimated that our share of the project would be $300,000,000 . . . and now when the project is 90 per cent completed we still stand by our original figure. This, gentlemen, is a great tribute to the engineering skills, and to the careful estimating and cost control of Ontario Hydro.

Celebration followed celebration in Cornwall and other St. Lawrence Valley communities during the next four days. July 2 was designated Ontario Day; July 3 was City of Cornwall and United Counties Day. On July 4, designated International Day, when Prime Minister John Diefenbaker visited the development, the great new lake was reaching its operating level and on the following day, at 6.39 p.m., the first unit of the Robert H. Saunders-St. Lawrence GS went into service.

There remained only the final scenes of the St. Lawrence drama: the ceremonial opening of the Seaway on June 26, 1959, in Montreal, and on the following day, the unveiling of the International Friendship Memorial at the power plants named after Robert H. Saunders and Robert Moses. At the Montreal ceremony, Her Majesty Queen Elizabeth II and President Eisenhower played complementary roles before a distinguished audience of Canadians and Americans. The great waterway, Queen Elizabeth said, which now could carry ocean shipping to the heart of the continent, would affect profoundly the lives of all future generations. In his turn, President Eisenhower spoke of the Seaway as a magnificent symbol to the entire world of how two democratic nations could work together peacefully for the common good, settling all differences "by patient and understanding negotiation, never by violence."

On June 27, when a heavy fog blotted out the river until almost noon, the Queen and her party first visited the Eisenhower Lock and Robert Moses Power Dam on the American side, then participated in unveiling ceremonies of the international boundary marker, and moved on to the reception centre of the Robert H. Saunders-St. Lawrence GS on the Canadian side. At the Eisenhower Lock they were greeted by Vice-President Richard M. Nixon, the Governor of New York, Nelson D. Rockefeller, and Robert Moses, Chairman of the New York State Power Authority; at the International Boundary by James S. Duncan, Chairman of Ontario Hydro, and at the Canadian reception centre by Leslie M. Frost, the Premier of Ontario. At the precise halfway mark of the continuous dam and power project, a 60-foot high parabolic arch of aluminum soars upward to span the boundary marker, and at each end of a walled exedra looking over the waters of the once-turbulent St. Lawrence, 88-foot flag poles, flying the Red Ensign of Canada and the Stars and Stripes, stand guard as it were beside the International Boundary Monument. It was to unveil this monument in connection with the official opening of the St. Lawrence Seaway that Queen Elizabeth had made her memorable journey to Canada. With guests of Ontario Hydro on one side and guests of the New York State Power Authority on the other, and after appropriate speeches by Chairman James S. Duncan, Robert Moses and Vice-President Nixon, the Queen pulled tasselled cords to disclose an immense slab of black granite whose polished surface bears in letters of bronze these inspiring words:

THIS STONE BEARS WITNESS TO THE COMMON PURPOSE OF TWO NATIONS, WHOSE FRONTIERS ARE THE FRONTIERS OF FRIENDSHIP, WHOSE WAYS ARE THE WAYS OF FREEDOM, AND WHOSE WORKS ARE THE WORKS OF PEACE

HYDRO IS STILL YOUNG

29

Less than sixty years ago Hydro was a struggling Cause, beset by the onslaughts of its enemies, its future dimly seen in the faint hopes of its friends. Today, its cause long won, Ontario Hydro is an established public electrical utility, much like any other except that it is bigger, covers more territory, and serves its customers —all of whom, directly or indirectly, are its owners—at cost. It has outlived the wrath of its enemies, the fears of its friends, and the effects of political convulsions which could have brought about its downfall. It has survived the dwindling of the undeveloped hydraulic power resources that gave it its name and reason for being. As a league of municipalities coöperating with and through a provincially appointed authority, and directly responsible to the people through their elected representatives, it has vindicated the long and bitter struggle for public ownership begun in Ontario at the birth of this century.

The technological and administrative accomplishments of Hydro have long attracted attention—far beyond the boundaries of the Province of Ontario, beyond Canada and the North American continent, and in fact, in every industrially progressive country in the world. The lengthy roster of foreign engineering experts who have visited Canada to study the structure and techniques of Hydro is in itself indicative of the lead taken in the intensive exploitation of electrical power potential. It is hardly by accident that the present Chairman of the Commission should himself be a dedicated internationalist, with a first-hand knowledge of most of the world's capitals.

A statistical comparison with other utilities shows that in general terms only the nationalized utilities of Great Britain and France, each serving a population more than seven times the size of Ontario's, are larger than Hydro. The actual service area of Hydro is more extensive than that of any other utility, with the possible exception of Russian systems. The Tennessee Valley Authority, for instance, while larger

268

in point of investment and installed generating capacity, operates within a much smaller area. Ontario Hydro's fixed assets, as represented by power plants and electrical facilities, exceed $2.5 billion—more than any other North American electrical utility.

In the province of Ontario the per capita consumption of electricity in 1959 was 5,950 kilowatt hours—among the highest in the world. This compares with the national figure of 5,700 for Canada as a whole, 4,500 for the United States, 2,100 (in 1958) for the United Kingdom, and 1,400 (in 1958) for France. As Mr. Duncan observed in an address to the Legislative Committee on Energy at Toronto early in 1960:

> It is interesting and significant to note, and it illustrates the advantages which the people of Ontario enjoy in relation to those living in certain less advanced countries, that the six million people living in Ontario consumed just about as many kilowatt hours of energy in 1959 as the 670 million people living in China.

Mr. Duncan summed up what Hydro means to Ontario in the following words:

> We are neither boastful that Hydro is among the largest of its kind in the world nor are we dismayed by the problems which an organization of these proportions presents.
>
> That the massive generation of low-cost power in this province has contributed in a major degree to the remarkable developments which have taken place in Ontario goes without saying.
>
> It is not by accident that the standard of living of our people has gone up 60 per cent during the last two decades, that the average per capita income is 18.5 per cent higher than the national average, that one out of every two immigrants to Canada elects to settle in Ontario, that 99 per cent of our homes have electric service and 50 per cent of Canada's industry is situated here.

At the end of 1959, electricity was available to 95 per cent of the farms in Ontario, including thousands of widely separated homesteads in remote northern areas. Considering the nature and extent of the terrain and the sparseness of its population—an average of ten customers for each mile of line, compared with an average of 120 for each mile of line in urban centres—this is a record of rural service unsurpassed anywhere. As the number of rural customers and their use of electricity increased, rates were reduced proportionately—from 3.35 cents per kilowatt hour in 1935 to 2.01 cents in 1959, while the number of customers increased over the same period from 67,802 to 491,100.

Impressive though this accomplishment may seem even when computed only in terms of inert statistics, the social and cultural values promoted by Ontario Hydro's rural developments have been infinitely more important. It is not generally recognized that the agricultural lands of Ontario comprise only 9.3 per cent of the province. Except for the clay belt of Northern Ontario, which still supports a boreal forest, most of the rural area of "Old Ontario" lies in the Pre-Cambrian Shield, the great peneplain of primordial rock whose southward-thrusting wedge splits Canada virtually asunder. On this inhospitable terrain as little as thirty years ago lived a population in conditions of economic, social, and cultural stagnation. The great stands of timber that brought it there, and supported it for a single generation, had succumbed to the bite of the lumberman's axe and the fires which followed inexorably in its wake. Inevitably, the wanton destruction of the forest values brought with it the deterioration of human values, which was reflected in the lassitude of the people, the inferior quality of the schools and the primitive character of the communications. The unpremeditated arrival on the scene of Model T Fords provided the first attack on these conditions—the improvement of roads and highways; but it was not until the appearance in the thirties of Hydro's steel-spurred construction crews to build their creosoted pole lines that the greatest civilizing boon of the Industrial Revolution—abundant electrical energy at negligible cost—was realized. Now, from the glacier-scarred outlines of the Shield north of Lake Ontario to the fast waters of the French River and beyond, there is hardly a farmhouse, and few summer cottages, without the immeasurable benefits of electricity. Bright lights replaced flickering kerosene lamps, and electric pumps, stoves and other labor-saving devices abolished the domestic drudgery of farm life, creating the incentive to modernize amenities and improve the appearance of the near-dilapidated houses. Equally important, initially, was the steady, well-paid work that the coming of Hydro meant for the chronically unemployed residents of these unproductive rural areas. Year-round employment on line-construction and maintenance helped disperse the lowering clouds of economic hardship, and together with Ontario Hydro loans to customers and Provincial grants in aid of construction, enabled the people to have electricity brought into their homes and workshops and to purchase electrical equipment and appliances. To anyone conversant with the so-called backwoods of Ontario, the miracle of social regeneration wrought by Hydro may well be regarded as its greatest triumph.

With almost all of the 1,528,000 dwellings in Ontario electrified, Hydro now covers every settled part of the province, through a network of 17,700 miles of transmission line and more than 47,000 miles of rural distribution line. Available power resources in 1959 reached 6.2 million kilowatts, sufficient to meet the increasing demand and provide a reserve for contingencies. In all, 820,000 kilowatts of new generating capacity were installed during the year, marking the greatest increase in generating capacity in any twelve-month period of the Commission's history.

Most of the new power came from the 940,000-kw Robert H. Saunders-St. Lawrence GS, while installation of a 200,000-kw unit at Richard L. Hearn GS brought its capacity to 600,000 kw. Scheduled for completion in 1961 by the addition of three more units, this station will provide a total of 1,200,000 kw to the Commission's power supply. In Northwestern Ontario a 45,500-kw station at Silver Falls, on the Kaministikwia River, was officially opened in September. Under construction and nearing completion were a 40,000-kw plant at Red Rock Falls on the Mississagi River and a plant at Otter Rapids on the Abitibi was scheduled to produce 90,000 kw in 1961 and a further 82,000 kw in 1963.

Construction work also continued on the thermal-electric stations at Lakeview on Lake Ontario west of Toronto and at Thunder Bay in the Northwestern Region, with first deliveries of power scheduled in both cases for 1961. The Lakeview station will have an ultimate capacity of 1,800,000 kw, equal to Hydro's total 1944 requirements. When completed, at an estimated cost of $250 million, the structure will measure 936 feet in length and will tower to the height of a twenty-storey skyscraper with stacks rising to 490 feet. Under peak conditions, its intake of cooling water will be at the rate of one million gallons a minute, more than five times the daily consumption of Metropolitan Toronto, and operating at full load it would burn 4.5 million tons of coal a year.

Determining general policy and maintaining control of the gigantic utility is the five-man Government-appointed Commission composed of Chairman James S. Duncan, First Vice-Chairman W. Ross Strike, Second Vice-Chairman Robert W. Macaulay, and Commissioners A. A. Kennedy and D. P. Cliff. Secretary of the Commission is E. B. Easson, who has the distinction of ranking, in terms of length of service with the Commission, second only to Mr. Strike.

W. Ross Strike celebrated his fifteenth year on the Commission in June 1959. The following year, in March 1960, at an OMEA-AMEU dinner held in his honor, he received this tribute from Mr. Duncan:

> It has been my privilege throughout the years to be closely associated with many distinguished and able men. In none of them have I found to a greater degree than in Ross Strike qualities of understanding, of friendship, of common sense, of sound judgment, of appreciation for the other man's point of view, and of readiness to defend the interests of the less privileged.

Over the past sixteen years, Mr. Strike has served under four chairmen and has achieved a record of service surpassed in length only by that of Sir Adam Beck. Originally and still a Commissioner of the Bowmanville Public Utilities Commission, Mr. Strike has been associated with this type of administration for more than twenty-eight years. A lawyer by profession, his knowledge of the municipal phases of the Hydro enterprise, and his familiarity with the Commission's operations, have made him an able counsellor both to the municipal utilities and to the Commission. His effective liaison between Ontario Hydro and the OMEA has tended to strengthen the bond which unites the Hydro enterprise.

In charge of administration and operation are J. M. Hambley, who succeeded A. W. Manby as General Manager on January 1, 1960; Dr. Otto Holden, Chief Engineer; together with five Assistant General Managers.

J. Mervyn Hambley has been with the Commission since 1930. He first joined Hydro as an Assistant Engineer in the Operating Department, serving the Georgian Bay System and Northern Ontario Properties.

In 1945 he was appointed District Operating Engineer for the Abitibi, Sudbury, Nipissing, Temiskaming and Manitoulin districts. Two years later, he was named Director of Operations for the Commission. In September 1953 he was appointed Deputy Assistant General Manager—Administration; in June 1959 Deputy General Manager, and on January 1, 1960, he became General Manager.

Born at Copper Cliff, Ontario, Mr. Hambley received primary and secondary school education at Copper Cliff and Sudbury. Employment for a year in the International Nickel Company smelter at Copper Cliff preceded four years at Queen's University, from which he graduated in 1929, with the degree of Bachelor of Science, Electrical Engineering. Following graduation he spent a year with the Canadian General Electric Company at Peterborough and Toronto before joining Ontario

Hydro. Always displaying a special interest in the human relations aspect of management, he was the Commission's first entrant in the University of Western Ontario's Management Training Course which he attended during the summer of 1952.

In 1959 Hydro employed approximately 15,900 persons, of whom 13,200 were on the regular staff. In its personnel and labor relations the Commission has for some years adopted the principle of collective bargaining, mainly with the Ontario Hydro Employees Union (OHEU). The wages, salaries and working conditions of most of the work force, with the exception of the engineering and supervisory classifications, are regulated by collective agreements periodically arrived at and paralleling closely the prevailing scales in business and industry.

The Ontario Hydro Employees Union was the new title adopted by the Employees' Association—formed as an independent union in 1945 —shortly after its affiliation in 1956 with the National Union of Public Service Employees, CLC (Canadian Labor Congress). It is the bargaining agent for the Commission's clerical, maintenance trades, operating staff and regular employees in the construction field forces and has approximately 10,000 members. In addition to the OHEU, the Commission has collective agreements with the International Union of Operating Engineers, certified by the Ontario Labor Relations Board to represent stationary engineers and trades employees in the two steam generating stations and the stationary engineers operating the heating plant in Head Office.

The Commission also pioneered a single collective agreement with the Allied Construction Council, comprised of construction craft unions who gained certification to represent Ontario Hydro's Construction Division employees at the Des Joachims GS and Otto Holden GS in 1950 and 1951. The Council was formed during the construction of the Sir Adam Beck-Niagara No. 2 project, which, because of its magnitude and complexity, created difficulties for Ontario Hydro in dealing separately with each of the eighteen construction craft unions involved. The Commission invited the unions to work out a satisfactory merger to bargain annually as one group through their representatives. The resulting merger, unique in Canadian history, proved so successful for both parties that the Allied Construction Council later gained representation of employees on all other projects built by Ontario Hydro throughout Ontario, as well as on its line and station construction.

When the St. Lawrence Power Project was begun in 1954, Ontario Hydro and the contractors pursued the joint bargaining policy by forming themselves into an organization, known as the St. Lawrence Labor

Relations Association. The association represented management in negotiating with the various unions engaged on the power project, and the construction employees were again represented by the Allied Construction Council.

About 960 of the 1,100 professional engineers employed by Ontario Hydro are represented by the Society of Ontario Hydro Professional Engineers, which is not a trade union.

Increasing wage rates and higher prices of equipment and materials, coupled with a continuing high level of construction (including entry into the nuclear-electric power field) and higher taxes and water rentals, have made an increase in the wholesale rates necessary in 1960. Because of the financial strength of the municipal utilities, retail rates to consumers will remain unchanged in most cases, as will rural retail rates with the exception of summer cottage contracts, where revenue falls far short of actual cost of service. As suburbanization proceeds, however, the prospect is that the rural system will lose more and more rural areas of high customer density to the municipal utilities through progressive annexation, thereby raising the unit cost of supplying power in the remaining thinly populated rural districts.

Ontario Hydro's rates, even should a general upward adjustment prove necessary in view of the steadily mounting production costs, will continue to be relatively low, since its customers will continue to enjoy the advantage, compared with most other utilities, of getting their electricity at cost. The principal problem facing the Commission is that of developing new power sources, now that the hydro-electric potential of Ontario has been almost fully developed.

When Hydro first began building its own generating stations the water-power resources of Ontario seemed inexhaustible. Half a century later, with the installation of the sixteenth unit at the Robert H. Saunders-St. Lawrence GS on December 18, 1959, there were no more hydro-electric sites to be developed except in the remote parts of Northern Ontario. It is estimated that by 1980 Hydro will have harnessed all the available residual sources of hydraulic power, adding a total capacity of between 1,500,000 and 2,000,000 kilowatts to the system.

Much of Hydro's future development, therefore, will depend on thermal and nuclear generated power. Already two of its coal-fueled thermal generating stations are in production, with the Lakeview and Thunder Bay stations scheduled for first delivery of power in 1961. These developments will ensure the availability of a power supply equal to the anticipated growth in Ontario's industrial and domestic requirements over the next few years; eventually, nuclear-generated electricity is expected to provide a long-term solution.

The Hydro-Electric Power Commission's *Brief to the Royal Commission on Canada's Economic Prospects,* presented in January 1956 by Dr. Hearn, contained this statement:

> Within the next twenty-five years . . . it is estimated that our total resources may be 23.6 million kilowatts. Of this amount 5.5 million kilowatts will be from hydro-electric generating stations, 10.6 million from conventional fuel-electric and almost 7.5 million from nuclear-electric generating sources.

In November 1958 these forecasts were endorsed and amplified in the Commission's *Brief to the Committee on the Organization of Government in Ontario,* presented by Mr. Duncan. This document, produced to provide the Committee with an outline of Hydro's development, the nature and scope of its operations, and its policies and practices, described Ontario Hydro's position regarding nuclear generation in some detail. Hydro's policy, it stated, is

> to develop the remaining (water-power) sites in conjunction with conventional thermal stations until nuclear-electric power is demonstrated to be approaching the cost of that produced by conventional plants. In the meantime, the coöperative effort (with Atomic Energy of Canada Limited and Canadian General Electric Company Limited) in the development of nuclear power will place Ontario Hydro in a position to build nuclear plants when it becomes advantageous to do so.

The experimental Nuclear Power Demonstration plant (NPD), located on the Ottawa River near Chalk River, is expected to have its 20,000-kw generating unit in service in 1961, when it will be maintained and operated by Hydro as part of its Southern Ontario System. It will be Canada's first nuclear-electric generating station. A larger project, the Douglas Point Nuclear Power Station on Lake Huron, is being planned as a full-scale generating station, with an initial capacity of 200,000 kilowatts in a single unit. Ontario Hydro is collaborating with AECL in the design and construction of the plant. Construction will begin in 1961. The total cost, exclusive of research expenditure, is estimated at $60,000,000. Initial supplies of power, scheduled for production in 1964, will be purchased by Hydro from AECL, and later, if it proves capable of generating electricity at costs competitive with those of conventional stations, the plant itself will be bought by Hydro.

In the meantime, notwithstanding the most intensive research and experimentation by physicists and engineers in many countries, no method has yet been discovered for the nuclear generation of energy at costs competitive with electricity derived from coal, gas, oil, or hydro-electric developments. It is for this reason that Ontario Hydro, having fully developed all the accessible hydraulic resources of the province, must now concentrate on the rapid construction of fuel-electric generating stations, the capacity of which now supplements, and will shortly exceed, the supply from hydro-electric stations.

Despite its long record of accomplishment, Hydro is still a young and growing enterprise, only now entering the period of its most rapid growth in the service of the Province to whose economy it has provided the principal dynamic. It has been able to do so because of the carefully wrought legislation that brought it into being; the integrity and ability of its several chairmen; the brilliant technical, engineering and administrative skills that were developed in its management, and because of the determination of the people of Ontario, as reflected in the attitudes of successive Provincial premiers, to keep it free from political influence or patronage. Though committed to the sale of energy at cost, Ontario Hydro from the outset has insisted upon the highest standards obtainable for equipment and construction and has operated with the goal of maximum security of service to its customers. A long-term result of these high standards has been relatively low operating and maintenance costs. In a large measure, however, Hydro's success has undoubtedly been due to the unique system of municipal coöperatives which served as the foundation for a public authority responsible to the Provincial Legislature but not under its direction or domination —in other words, to those very principles of public ownership that were so virulently attacked as being inimical to efficient and honest administration. Federal and Provincial Governments in Canada have, of necessity, subsidized or have undertaken directly the development of transportation, communications, and many natural resources. Among such enterprises, Ontario Hydro has become an outstanding example of successful public ownership and management without requiring any financial aid from Government or taxpayers other than the capital grants for rural electrification. This is to be seen not only in the operations of Ontario Hydro as the senior authority in a complex integration of coöperatives, but in those of the municipal commissions, which not only manage their local utilities but have a voice also as "stockholders" in the formulation of Hydro policy. In both capacities, the municipalities, through the OMEA and AMEU, have exerted a far greater and more beneficent influence on Hydro's development than is generally realized.

Fifty years ago, when Hydro was struggling to establish and prove itself, its opponents declared that it must fail because no publicly owned utility could compete successfully with the profit-motivated drive and efficiency of a private corporation. History has proved them wrong. The splendid dream of Detweiler and Snider, the embattled Cause of Adam Beck, has become one of the world's largest and most successful organizations engaged in the production and distribution of electrical energy. It has survived and succeeded because, in all its departments and functions, Hydro has enjoyed a decisive advantage that stems from, and will forever remain rooted in, its profound and implicit belief in the motto it so proudly displays: *Dona Naturae Pro Populo Sunt*—the Gifts of Nature are for the People.

Appendix 1

APPOINTMENT	CHAIRMEN	RETIREMENT	COMMISSIONERS AND VICE-CHAIRMEN	RETIRE- MENT
June 1906	Adam Beck	Aug. 1925	J. S. Hendrie	Sep. 1914
			C. B. Smith	Feb. 1907
Aug. 1906				
Oct. 1906				
Feb. 1907			W. K. McNaught	Feb. 1919
Oct. 1909				
Dec. 1912				
Oct. 1914			I. B. Lucas	July 1921
Nov. 1919			D. Carmichael	June 1923
July 1921			F. R. Miller	Aug. 1922
Jan. 1923			J. G. Ramsden	July 1923
July 1923			J. R. Cooke	June 1931
Sep. 1925	C. A. Magrath	Feb. 1931	C. A. Maguire	July 1934
June 1931	J. R. Cooke	July 1934	A. Meighen	May 1934
July 1934	T. S. Lyon	Oct. 1937	A. W. Roebuck	Apr. 1937
			T. B. McQuesten	Oct. 1937
Feb. 1936				
Nov. 1937	T. H. Hogg (*and Chief Engineer*)	Feb. 1947	W. L. Houck	Aug. 1943
			J. A. Smith	Aug. 1943
Nov. 1938				
Aug. 1943			G. H. Challies	May 1955
June 1944			W. R. Strike	
Mar. 1947				
Jan. 1948				
Mar. 1948	R. H. Saunders	Jan. 1955		
Jan. 1955	R. L. Hearn	Oct. 1956		
May 1955			W. E. Hamilton	Aug. 1955
			A. A. Kennedy	
Aug. 1955			W. K. Warrender	Oct. 1956
Nov. 1956	J. S. Duncan		T. R. Connell	May 1958
			D. P. Cliff	
May 1958			R. W. Macaulay	
Jan. 1960				

SECRETARIES	RETIREMENT	CHIEF EXECUTIVES	RETIREMENT
		P. W. Sothman (*Chief Engineer*)	July 1912
E. C. Settell	Sep. 1909		
W. W. Pope	Jan. 1936		
		F. A. Gaby (*Chief Engineer*)	July 1934
		T. H. Hogg (*Chief Engineer-Hydraulic*)	Oct. 1937
A. M. McCrimmon (*Secretary & Controller*)	Nov. 1937		
		T. H. Hogg (*Chief Engineer*)	Feb. 1947
O. Mitchell	Dec. 1947		
		R. L. Hearn (*General Manager & Chief Engineer*)	Jan. 1955
E. B. Easson			
		A. W. Manby (*General Manager*)	Dec. 1959
		O. Holden (*Chief Engineer*)	July 1960
		J. M. Hambley (*General Manager*)	

Appendix 2

Names of the Chairmen and Senior Operating Officials of the Fourteen Municipal Commissions which Entered into Contracts with Ontario Hydro in May, 1908.

MUNICIPALITY	CHAIRMAN	SENIOR OPERATING OFFICIAL
Berlin	A. L. Breithaupt	E. J. Philip, *superintendent*
Galt	F. S. Scott	Robert Elliot, *superintendent*
Guelph	Samuel Carter	J. J. Heeg, *superintendent and secretary*
Hespeler	Peter Jardine	M. E. Jardine, *secretary-treasurer*
Ingersoll	George Sutherland	Harold Hall, *manager*
London	Philip Pocock	F. R. Dark, *superintendent*
New Hamburg	John Hesse	George Morley, *superintendent*
Preston	F. Clare	T. R. Waugh, *superintendent*
St. Mary's	John Willard	E. J. Stapleton, *superintendent*
St. Thomas	W. K. Sanderson	E. H. Caughell, *superintendent*
Stratford	J. J. Mason	R. H. Myers, *secretary-treasurer*
Toronto	P. W. Ellis	H. H. Couzens, *manager*
Waterloo	Aloyes Bauer	Ford S. Kumpf, *manager*
Woodstock	D. W. Karn	J. G. Archibald, *manager*

Appendix 3

The Hydro-Electric Power Commission of Ontario
STATISTICAL SUMMARY
ALL SYSTEMS COMBINED

Power Resources

		1919	1939	1959
December Dependable Peak Capacity of:				
Power Purchased	KILOWATTS	62,725	499,800	620,900
Commission Generating Stations	KILOWATTS	182,400	1,058,700	5,533,800
Total	KILOWATTS	245,125	1,558,500	6,154,700

Power Demands

December Primary Peak Requirements	KILOWATTS	243,100	1,317,000	5,556,500

Annual Energy Production

	MILLIONS OF KILOWATTHOURS			
Purchased		N.A.	2,763	5,865
Generated		N.A.	5,910	29,600
Total		N.A.	8,673	35,465

		1919	1939	1959
Transmission Lines	CIRCUIT MILES	2,858	8,838	17,713
Rural Distribution Lines	MILES	N.A.	17,706	47,351
Net Revenue	MILLIONS OF DOLLARS	3.8	36.6	212.6
Total Assets	MILLIONS OF DOLLARS	59.5	363.6	2,548.3
Customers served directly or indirectly through Municipal Electrical Utilities	NUMBER	234,426	721,997	1,830,453

Average Energy Consumption per Customer

Domestic Service—Municipal Systems	KILOWATTHOURS	404	2,000	5,474
Farm Service	KILOWATTHOURS	N.A.	1,652	5,718

N.A.—Not Available.

Glossary

Some Practical Electrical Terms

ALTERNATING CURRENT An electrical current which alternately flows in opposite directions in an electric circuit.

AMPERE Unit by which the flow of current is measured. Named for Andre Marie Ampere, 1775-1836, French physicist, mathematician and philosopher.

c.f.s. (CUBIC FEET PER SECOND) Unit by which the volume of flowing water is measured.

CYCLES The number of times per second alternating current completely reverses the direction of its flow.

DIRECT CURRENT An electric current which always flows in one direction in an electric circuit.

FOREBAY A pond or reservoir of water from which water is supplied through penstocks to the hydraulic turbines.

GENERATOR A machine for converting mechanical power into electrical power.

GS Generating station.

hp (HORSEPOWER) A rate of doing work. An Electrical Horsepower equals 746 watts or approximately three-quarters of a kilowatt.

HYDRAULIC TURBINE An enclosed rotary prime mover which produces mechanical power by utilizing the energy given up by water flowing from a high elevation to a lower elevation.

kv (KILOVOLT) 1,000 volts.

kw (KILOWATT) 1,000 watts.

kwh (KILOWATT-HOUR) Basic unit of electric energy equivalent to one hour's use of 1,000 watts or one kilowatt. For example, a 100-watt lamp burning for 10 hours.

PENSTOCK A pipe for delivering water to a hydraulic turbine.

TRANSFORMER An apparatus for changing the voltage of an alternating current supply.

TS Transformer station.

VOLT Unit of electrical pressure which makes the current flow. Named for Alexandra Volta, 1745-1827, Italian physicist.

WATT Electrical unit of power or rate of doing work. Named for James Watt, 1736-1819, Scottish engineer and inventor.

INDEX